THE GLOBAL PIGEON

FIELDWORK ENCOUNTERS AND DISCOVERIES

A Series Edited by Robert Emerson & Jack Katz

THE GLOBAL PIGEON

Colin Jerolmack

THE UNIVERSITY OF CHICAGO PRESS
Chicago & London

COLIN JEROLMACK is assistant professor of sociology and environmental studies at New York University.

The University of Chicago Press, Chicago 60637
The University of Chicago Press, Ltd., London
© 2013 by The University of Chicago
All rights reserved. Published 2013.
Printed in the United States of America

22 21 20 19 18 17 16 15 14 13   1 2 3 4 5

ISBN-13: 978-0-226-00189-0 (cloth)
ISBN-13: 978-0-226-00208-8 (paper)
ISBN-13: 978-0-226-00192-0 (e-book)

Library of Congress Cataloging-in-Publication Data

Jerolmack, Colin.
  The global pigeon / Colin Jerolmack.
    pages ; cm. — (Fieldwork encounters and discoveries)
  ISBN 978-0-226-00189-0 (cloth : alkaline paper)
  ISBN 978-0-226-00208-8 (paperback : alkaline paper)
  ISBN 978-0-226-00192-0 (e-book)
  1. Feral pigeons—Social aspects. 2. Feral pigeons—Habitat. 3. Human-animal relationships—Social aspects. 4. Feral pigeons—Social aspects—New York (State)—New York. 5. Feral pigeons—Habitat—New York (State)—New York. 6. Human-animal relationships—Social aspects—New York (State)—New York. I. Title. II. Series: Fieldwork encounters and discoveries.
QL696.C63J47 2012
598.6'5—dc23

2012017790

♾ This paper meets the requirements of
ANSI/NISO Z39.48-1992 (Permanence of Paper).

For DOUGLAS PORPORA,

friend, mentor, and intellectual hero

> Believing that it is always best to study some special group, I have, after deliberation, taken up pigeons.
>
> CHARLES DARWIN

CONTENTS

INTRODUCTION
Experiencing the City through the Quintessential Urban Bird
1

I. The Pedestrian Pigeon

ONE
Feeding the Pigeons
Sidewalk Sociability in Greenwich Village
23

TWO
"Do Not Feed the Pigeons"
Cultural Heritage and the Politics of Place in Venice and London
44

II. The Totemic Pigeon

THREE
New York's Rooftop Pigeon Flyers
Crafting Nature and Anchoring the Self
79

FOUR
The Turkish Pigeon Caretakers of Berlin
Primordial Ties in a Migrant Community
109

FIVE
Joey's Brooklyn Pet Shop
Cosmopolitan Ties in a Changing Urban Landscape
133

III. Deep Play

SIX
The Bronx Homing Pigeon Club
Nature, Nurture, and the Enchantment of "the Poor Man's Horse Racing"
159

SEVEN
South Africa's Million Dollar Pigeon Race
Rationalizing and Globalizing "the Pigeon Game"
192

EIGHT
Conclusion
Changing Ecologies
221

Acknowledgments 239
Notes 243
References 257
Index 269

INTRODUCTION

Experiencing the City through the Quintessential Urban Bird

I NEVER PAID MUCH attention to pigeons until one defecated on me. During my years as a graduate student with an interest in public space, I spent many days wandering around the streets of Greenwich Village in Manhattan—an area famously consecrated as a model urban community in Jane Jacobs's 1961 book *The Death and Life of Great American Cities*. I found one small space called Father Demo Square particularly intriguing because, physically, it was a dingy, tiny traffic island; but, socially, it was a hub of activity in the heart of the Village. On my first visit to Father Demo Square in February 2004, over a dozen pigeons were perched on top of a cluster of three lamps in the middle of the space, directly above a sign hung by the Parks Department that read "DO NOT FEED THE PIGEONS." As I leaned in to read the faded dedication to Father Demo that was etched into a marble base under the lamps, a pigeon atop the light dropped its fecal load. The green-white liquid glanced off the side of my head and splashed on the pavement in front of me. Indeed, the pigeons were out in droves. Several dozen marched around the central light fixture, cooing and flapping and pecking for scraps like chickens in a yard.

In my frequent visitations to the park over the next year, I noticed that dozens of people fed pigeons there every day. Many fed them spontaneously with small pieces of their sandwiches, but a handful of people poured bags of grain for the birds at specified times. The pigeons were habituated to humans because of these routines, able to coax people into tossing food by approaching them and cocking their heads. Pigeons seemed to recognize the regular feeders, following them into the park and even climbing

on their laps and arms. It was also common for children to chase the pigeons and for visitors to photograph them.

Despite the relative popularity of pigeon feeding, some visitors avoided the feces-splattered benches or complained about the "pigeon lady," a well-known elderly white woman named Anna, who dumped feed for the birds three times each day. There was also institutional opposition to pigeon feeding. In the communications of local citizen groups, the community newspaper, and the Parks Department, pigeons kept coming up as one of the two most pressing problems perceived to be plaguing Father Demo Square. Locals hoped that a new fence and the imposition of a curfew after scheduled renovations would prevent homeless people from spending the night there, but the "pigeon problem" seemed intractable. Simply put, the square was a small space with a lot of pigeons that stayed put because a sizable number of people fed them. The head of a local block association lamented that the park had become "a mini Piazza San Marco," a reference to the famously pigeon-inundated Venetian plaza.

Through Father Demo Square I saw two significant but contradictory ways that pigeons figured in how people interpreted and experienced urban space. For those who fed them, pigeons provided casual and spontaneous opportunities for playful sidewalk interaction; they were like a public amenity, enhancing people's enjoyment of the place. But for those who disdained the presence of pigeons, these animals were what anthropologist Mary Douglas called "matter out of place."[1] They did not *belong* in the park. It was no coincidence that the problems posed by pigeons in the square were conflated with the problems posed by homeless people. In quite explicit ways, including a report by a Brooklyn councilperson titled "Curbing the Pigeon Conundrum," pigeons and pigeon feeders in New York were framed as a "quality of life" issue whose presence threatened a sense of civility, order, cleanliness, and safety in public space.

Despite the fact that the city of New York classified pigeons as "nuisance animals," in the section of Brooklyn where I lived, there were over a dozen men who hand-raised hundreds of these birds on their rooftops and boldly commanded them to fly in massive bundles across the sky like a swarm of locusts. From the sidewalks of the blue-collar but gentrifying neighborhood of Bushwick, I curiously watched their masses of pigeons hover over the hectic streets and glitter in the sun's golden hour before dusk. In 2005, I started making regular visits to a nearby pet shop where "pigeon flyers" congregated. Over the next three years, I came to know 44 working-class men who kept pigeons in rooftop "lofts"[2] throughout Brooklyn and Queens.

The first pigeon flyer to invite me to his roof was a sarcastic but kindhearted old-timer named Carmine Gangone, a lifelong "pigeon guy." Even at 86 years old, he ascended the rung ladder to the roof of his townhouse in Ozone Park, Queens, at dawn to chase his 150 or so domesticated pigeons out of their shacklike coop and into the sky. Waving a long bamboo pole, Carmine would startle the birds into flying higher together in a

Photo by Marcin Szczepanski.

coordinated, tight bundle that he referred to as a "stock." He delighted in the hooks, turns, and spirals that the stock made as a unit; and, if he was lucky, his birds might return with a "stray" pigeon that joined them from a distant flyer's stock. Once the birds were airborne, Carmine would scrape the feces from his coop using a hoe, change the water, and check on the babies he was raising. A retired plane cargo loadmaster, a widower, and a World War II vet, Carmine treated the hobby like a job. For one of the few Italian Americans who remained in a neighborhood that had become a destination for Latino and Southeast Asian migrants, the pigeon coop gave Carmine a sense of constancy.

From afar, I romantically imagined the pigeon coop as a modern-day urban manifestation of Henry Thoreau's cabin on Walden Pond, enabling the flyers to escape the concrete jungle and find solace in intimate relations with the "natural world." But Carmine and the other flyers disabused me of such fanciful notions. Rather than seeking communion with nature, Carmine absconded to his rooftop with a bit of nature's raw material and relished his power to sculpt the pigeons, through selective breeding and training, according to his will. Yet Carmine's tastes were hardly idiosyncratic. He crafted his "purebreds" and judged their worth based on the social standards of his peer group. His birds' performance and appearance were also the resources with which Carmine competed for status among the flyers.

The flyers' relations with their birds augmented connections to humanity more than nature. It was *through* pigeon flying that Carmine found his place in the urban community and constructed what sociologist George Herbert Mead called the "social self." While Carmine waxed nostalgic about the loss of the city's idyllic Italian neighborhoods, his coop provided a palpable link to that past while also enabling him to build new social ties with some of the men of color who replaced his coethnic peers. Though pigeon flying—an iconic and once popular New York pastime that was historically the domain of ethnic whites—is disappearing, Carmine shared the hobby with Hispanic and black men who picked up pigeon flying decades ago from ethnic whites after moving into neighborhoods in transition such as Bushwick. Through their collective involvement in rooftop pigeon flying and pet shop socializing, these men built a social world with its own unique set of pleasures, rules, and rewards.

This is a book about how interactions with animals—pigeons, in particular—animate people's social worlds and their experience of the city. In Father Demo Square and nearby parks, encounters with pigeons influenced people's behavior in, and interpretation of, public space. Up on the roof, the flyers' relationships with their birds shaped their friendship networks, their experience of their neighborhoods, and even their sense of self. "Urbanism as a way of life" is typically construed as a human existence devoid of encounters with animals and "organic nature," and social scientists usually exclude nonhumans from the social realm.[3] But the main point of this book is to illustrate how cross-species encounters can in fact be a constitutive feature of social life in the city. What do such encounters mean in people's everyday lives, and how are these meanings patterned by context, class, race, gender, and place? Why do people seek relationships with other species? What does the place that people grant or deny animals in their built environments reveal about how they experience and imagine the city? I explore these questions by drawing on extended participant observation in New York as well as abbreviated ethnographic case studies in Berlin, London, Venice, and Sun City (South Africa).

Though I did not set out to write a book about pigeons, I realized that Father Demo Square and the rooftop flyers gave me a unique window into the changing character of urban neighborhoods, street life, and city dwellers' relationship to the "natural world." I saw that pigeons were one of a growing number of species targeted for removal as part of municipal efforts to "clean up" New York. And I learned that the waning of rooftop coops was tied to a vanishing way of life rooted in crowded tenements and ethnic enclaves, and that ethnic whites passed on their hobby to nonwhites as their neighborhoods turned over. Curious about the "classed" and parochial character of pigeon flying, I also spent three years observing a pigeon racing club in the Bronx. Long a staple of the urban proletariat in Europe and the United States, pigeon racing is a sanctioned sport in which contestants compete against their neighbors through local clubs. Though Bronx Homing Pigeon Club members were men of modest means living in cramped quarters, with pigeons they could breed "thoroughbreds" and put on their own Kentucky Derby. But I saw that the sport's allure went beyond cash or status: through their birds, so-called pigeon fanciers vicariously engaged in a dramatic struggle against the powerful and hostile forces of nature—without leaving their roof. The return of their pigeon from a 400-mile marathon was a magical, miraculous triumph of nurture over nature and a validation of the labor they poured into training.

I decided to expand my research beyond New York when, during a visit to Berlin in 2005, I serendipitously met a group of Turkish men who kept pigeons. Over the next month, I saw how, by importing an animal practice that was common to their homeland, these immigrants used the pigeon coop to carve out a tangible "Turkish space" and express their ethnicity in a foreign land. Comparing the New York and Berlin flyers made clearer to me how their practices, and the meanings they ascribed to them, were patterned by their social situation. I found it fascinating that, while pigeon keeping was a means for the New York flyers to transcend race and ethnicity, this practice reinforced ethnic attachments in Berlin. Despite these differences, for both groups the pigeon coop was simultaneously an embodiment of the men's nostalgia for a lost world and an organizing principle of their present-day social relations and identity.

My experience in Berlin inspired me to identify additional strategic research sites that I could use as a point of analytical comparison for my observations in New York. To further elucidate the contested place of pigeons on city sidewalks, I spent the month of June 2006 in Trafalgar Square in London and Piazza San Marco in Venice. Until recently, both sites famously hosted vendors who sold pigeon feed. It was a tradition for visitors to allow the ravenous flocks of pigeons to land on their shoulders and eat from their hands. I saw that this ritual led pedestrians to interpret the pigeon flocks as rightful—even celebrated—residents of the squares, and that the decision in both cities to evict the seed vendors, redefine pigeons as "rats with wings," and ban pigeon feeding was part of a larger political project aimed at tackling a host of "quality of life" issues. Juxtaposing Trafalgar Square, Piazza San Marco, and Father Demo Square made salient for me the ways that encounters with pigeons mediated people's experience of urban spaces.

Finally, my research in the Bronx revealed that the sport that fanciers lovingly call "the poor man's horse racing" was being professionalized: fanciers were increasingly racing for large cash prizes, not trophies; breeders were selling their pedigreed "studs" for hundreds of dollars; and homers were being tested for drugs. I even heard of fanciers bypassing their local club races to compete in new high-stakes international races. In February 2006, I "followed the money" to the sport's marquee event in Sun City, South Africa: the Million Dollar Pigeon Race (MDPR). Unlike traditional races, fanciers did not train their pigeons, compete against neighbors, or wait at their lofts for their birds' return. Instead, fanciers around the globe shipped their best birds to Sun City, where they were conditioned by paid experts. On race day, I sat among thousands of gambling spectators in Sun City's luxurious casino and watched, via a jumbo video screen, as the winning pigeon touched down. It netted its owner and breeder $120,000 and was auctioned the next day for $12,000. In delocalizing pigeon racing and contracting out its myriad requisite tasks to an array of professionals, I saw how the MDPR modeled ideals of global inclusiveness and a "level playing field." However, this rationalized version of the sport deskilled fancying, priced out the working class, and circumscribed the possibility that hard work could trump big pockets.

City Birds

The pigeon is a felicitous animal for exploring the social significance of cross-species urban entanglements. These birds are thoroughly entrenched in the cityscape. As ground feeders and the descendents of cliff dwellers, pigeons are right at home pacing the sidewalks to scrounge for food and making nests on window ledges. They stalk among commuters, coax park visitors into feeding them, and scavenge the waste that humans leave behind. Although urbanization endangers many plant and animal species, pigeons are exquisitely adapted to a brave new world where ecosystems are laid over concrete. Despite the fact that people may treat the sidewalk as a "defended territory" in which these unmanageable creatures are unwelcome "invaders,"[4] pigeons have in effect become naturalized urban citizens. Their presence on city streets is utterly *pedestrian*, in both senses of the word. It was this pedestrian quality of pigeons—mundane, commonplace streetwalkers—that brought them into my sociological field of vision as I observed everyday urban life in Greenwich Village.

Commonly referred to as "rats with wings," a label meant to characterize them as filthy vectors of disease, pigeons in New York—and many cities worldwide—are typically considered pernicious "nuisance animals." New York's built environment includes myriad (often futile) devices designed to repel pigeons: plastic owls on rooftops; spikes on ledges and over doors; speakers blaring recordings of birds of prey from the cornices of Union Square, Times Square, and the JFK airport; and the ubiquitous green "Do Not Feed the Pigeons" signs that the Parks Department has placed in nearly every square and park. In 2003, Bryant Park even employed a trained hawk to scare away the pigeons, though it was decommissioned after mauling a pedestrian's Chihuahua. And in 2007, a Brooklyn councilman's report recommended the appointment of a "pigeon czar" to oversee pigeon control and a $1,000 fine for feeding pigeons. The city also considered feeding its pigeons birth control, a method adopted in Hollywood, and purchasing $4,000 robotic hawks, used in Liverpool. Pigeons—rugged, unafraid, and adaptable—seem to be the quintessential city bird, but one that many urbanites wish would disappear. This incongruity was on display when the New York City Parks Department selected "Parker the Pigeon" as a finalist for the "perfect" New York mascot even though the Parks Department was actively involved in controlling pigeons and regarded them as a nuisance.[5]

New York City is hardly unique in its efforts to control pigeons. Across North America and Europe, and in many cities in Asia, Africa, and South America, antipigeon artifacts—such as sticky gel strips on ledges and nets over door entrances—are ensconced features of urban architecture. Cities and towns around the world have also criminalized pigeon feeding to control their numbers and the problems linked to them, from respiratory diseases that are potentially transmissible to humans, such as psittacosis, to the property damage that can result from their acidic feces. Though the threat of disease is one of the most often cited rationales for pigeon control (and pigeon loathing), epidemiologists and biologists consider the public health risk of street pigeons to be "very low, even for humans involved in occupations that bring them into close contact with nesting sites."[6] Their threat to property, however, is more tangible.

One pigeon can produce up to 25 pounds of droppings per year, and a 2008 study of Venice, Italy, claimed that the city—with a pigeon population estimated to be as high as 130,000—spent an annual average of €16–23 *per pigeon* to remove the guano from its historic monuments and streets. London's former mayor Ken Livingstone calculated that it cost $235,000 every year to clean up pigeon feces in Trafalgar Square; and one estimate put the annual cost of pigeon-related damage to property in the United States at $1.1 billion. In response, from Pittsburgh to Prague control tactics include shooting, electrocuting, and poisoning pigeons.[7]

People have not always deemed pigeons "nuisance animals" or sought to reduce their numbers. And it is only in the last century or so that pigeons have come to be considered distinct from doves. "Rock pigeons," also called "rock doves" (*Columba livia*), were first domesticated about 5,000 years ago, and for millennia humans selectively bred them to meet a variety of material needs. The legacy of these efforts is omnipresent: the gray pigeons with black-barred wings and iridescent neck feathers that occupy city streets worldwide today are rock pigeons' feral descendents. Their origin is usually traced to North Africa, parts of Mediterranean Europe, the Indian Subcontinent, and central Asia; but commerce brought them around the world.

Pigeons partly domesticated themselves. It is thought that the mud and stone walls of ancient human dwellings, so similar to the rock pigeon's natural habitat of rocky ledges, cliffs, and caves, served as decent nesting sites. Pigeons were also attracted to the grains that emerging agricultural societies produced. Rather than viewing these avian interlopers as pests, humans saw them as a source of food, and they discovered that pigeon guano made excellent fertilizer. Rock pigeons were valued and even revered in premodern times for their "reproductive magic," breeding more often and for a longer season (year round) than most other animals. Their gregariousness and docility also made them fitting symbols of peace (their predators, hawks, stand for aggression).[8] Humans took advantage of pigeons' adaptability and submissiveness, building houses called dovecotes that lured the birds into semidomesticity and allowed for the easy harvesting of their flesh, eggs, and feces. In feudal times, pigeon meat and guano were deemed so valuable that the right to own a dovecote was restricted to nobility—no wonder, then, that so many of these ornate structures were toppled in the wake of the French Revolution.⁹

Rock pigeons are extraordinarily malleable. Over time, humans bred larger and fatter varieties for food, and stronger and leaner varieties with enhanced "homing" instincts (called homers) that could carry messages over hundreds of miles. But many pigeon breeds were created for non-instrumental ends; they instead reflected the aesthetic and leisurely preferences of fanciers (breeders), who by the 19th century had organized clubs and competitions based on the appearance or performance of particular pigeon breeds—akin to the Westminster dog show or greyhound racing. Long-distance pigeon races became particularly popular in Belgium and other parts of western Europe. By the early 20th century, the sport had been imported to the United States. New York alone boasted dozens of homing pigeon racing clubs; and rooftop and backyard coops became common features of industrial cityscapes, from Chicago to Germany's Ruhr district.

The historian James Secord attributes the popularity of pigeon fancying in Western cities to the fact that it "provided the harried urban dweller with a link, however tenuous, to a rural and Arcadian past" and served as a pretext for "social gatherings and congenial conversation." Pigeons were cheap, easy to breed, and needed little space or attention. Further, Secord speculates that fancying "fascinated the Victorians by keeping nature close at hand yet under control."¹⁰ Pigeon fancying was so popular, and the variation in shapes, sizes, and colors that fanciers had produced through centuries of selective breeding was so extraordinary, that Charles Darwin opened *On the Origin of Species* with an exhaustive genealogy of so-called "toy" pigeon breeds. The "selecting hand" of nature "was manifested . . . in man's actions as a breeder." Skeptical of his scientific theory, Darwin's colleagues had urged him to scrap *Origin* in favor of a book solely about pigeon breeds. His editor remarked, "Everybody is interested in pigeons."¹¹

The pigeons that occupy our sidewalks never existed in the wild. They are descendants of escaped domesticated pigeons that were imported to the United States, Europe, and elsewhere centuries ago. In fact, it was French settlers who introduced the rock dove to North America in the early 1600s, primarily for consumption. With humans' help, pigeons have proved to be uniquely adept at living in urban settings, replacing cliffs with cornices. But pigeons are no longer useful to most societies as messengers or as sources of food and fertilizer, having been replaced by

A fantail and a pouter, two of Darwin's favorite "toy" breeds. Photos by Jim Gifford.

cheaper and more efficient alternatives. And pigeon fancying is far less popular than in the Victorian era, so much so that the intimate associations that fanciers have with their birds strike some observers as anachronistic. Finally, though in decades past it was so popular to feed street pigeons that urban parks commonly had dedicated feeding areas and even vendors or machines selling seed for the birds, these days pigeons—and the people who feed them—are often unwelcome in public spaces. The pigeon's fecundity, nitrogen-rich feces, proclivity to return home, and easy adaptability to human environments—traits that people once valued and intentionally enhanced—are exactly the traits that bother so many contemporary urban dwellers.

Society, then, has abetted an animal whose niche is one designed to be the exclusive habitat of humans: the sidewalk. Having forsaken pastoral life, pigeons humble our efforts to place a firewall between "urban" and "natural." These nonnative, feral birds—neither purely wild nor domestic—now confront humans as our own historical detritus and are regularly met with public antipathy. But because pigeons breed up to the available food supply, efforts to control them are doomed to fail as long as there is available food, which cities provide in abundance thanks to their prolific amounts of discarded organic garbage and to the many people who intentionally feed them. The result is that pigeons seem to be inextricably enmeshed in a symbiotic—some say parasitic—relationship with society. They are what biologists call synanthropes (literally, "together with man"), preferring to live among people in the built environment.[12] In true Darwinian fashion, pigeons have adapted their habits to our own.

Animals, Nature, and Urbanism

Synanthropes like pigeons challenge the conventional notion that urbanization has insulated people from contact with animals and nature. In a classic essay titled "Why Look at Animals?," the art and literary critic John Berger argues that, while people lived close to animals and regarded them with respect and even awe in "traditional" societies, urbanization has resulted in the "disappearance of animals from daily life" and in their debasement. Berger contends that zoos, and even pets, represent "monuments to the impossibility" of meaningful encounters with animals *as animals*; the animals we keep in cages and in our homes have been thoroughly denatured and humanized.[13] Following Berger, many scholars and environmentalists situate humanity's assumed estrangement from animals within a larger narrative about urbanized society's alienation from nature. The biologist Edward O. Wilson has famously hypothesized that humans have an instinctive attraction and need for close contact with nature, but that the *biophilia* impulse has atrophied as "artificial new environments" saturate our lived experience.[14] The journalist Richard Louv provocatively diagnosed Americans as suffering from "nature deficit disorder," and ecologist Robert Michael Pyle decries the "extinction of experience" that marks modern living: "As cities and metastasizing suburbs forsake their natural diversity, and their citizens grow more removed from personal contact with nature, awareness and appreciation retreat."[15] This societal alienation from nature, in turn, is seen as a foundational cause of our contemporary ecological crisis.[16] I call this argument "Nature Lost," because it parallels sociology's classic "Community Lost" thesis in which the same culprit—urbanization—is said to undermine traditional forms of communal life and produce social disorganization.[17]

If the Nature Lost thesis is correct, then fostering environmental consciousness entails reinvesting the nonhuman world, including the ordinary species around us, with transcendental meaning. Pyle tells urbanites to discover "Eden in a vacant lot" and wonders if the vanishing of the condor could be at all meaningful to a child who never learned to see the beauty of nature in the common birds in her neighborhood. In surveying Americans' responses to the decline of open space and biodiversity, sociologist James Gibson argues that a "culture of enchantment" is in fact being reawakened in the collective conscience. He finds a growing number of people today "who long to rediscover and embrace nature's mystery and grandeur . . . and who look to nature for psychic regeneration and renewal."[18] New Yorkers celebrated when hawks built a nest on Fifth Avenue, and a Colorado town grieved when a friendly elk was poached, because people felt a sacred, primordial bond with the natural world through these animals. This ethos starkly contrasts with the mechanistic and utilitarian view of animals and landscapes that is said to be part and parcel of the rationalization of society through science and capitalism, which classical sociologist Max Weber memorably claimed, "drained the cosmos of magic."[19] Indeed, Gibson sees a clear parallel between the "culture of enchantment," personified in but not limited to the environmental movement, and the totemic cosmologies of many premodern indigenous societies, which he says worshipped nature and believed that humans and nonhumans were kin.

While the Nature Lost thesis engenders an important critique of contemporary Western environment-society relations and captures a popular way that people interpret nature, it implies that relations with nonhumans can—and ought to—be "free from social interests"[20] and that people's desire for relationships with other species is driven by a singular deep-seated need to connect to nature. To commune with other species is to transcend society. But this *asocial* conception of nature—epitomized in the mystical prose of Henry Thoreau and John Muir—is, as environmental historian William Cronon has shown, a modern "cultural invention" that reflects urbanized society's rose-colored nostalgia for simpler times. "The way we interpret nature," Cronon adds, "is so entangled with our own . . . cultural assumptions that the two can never be separated."[21] In the same vein, environmental sociologist Michael Bell's nuanced ethnography of an English exurban village revealed that residents' sanctification of nature and self-identification as "country people" were grounded in their quixotic search for a "moral preserve in a landscape of materialist desire" and class conflict. But, belying their belief that nature offered sanctuary from social life, Bell found that villagers related to the environment differently depending on their class position.[22]

In my research, I did not view cross-species encounters as alienated or inauthentic if they did not foster a sense of connection to nature, nor did I assume that close relations with animals resulted from an internal human longing to commune with the nonhuman world. As an ethnographer, I sought to comprehend these interactions from the actor's point of view. As a sociologist, I aimed to grasp how people's interpretations of cross-species encounters—whether marked by feelings of transcendence, affection, disgust, or something else—were patterned by the *social contexts* in which they were embedded. While this book builds on scholarship that exposes the social machinery lurking under people's ostensibly asocial experience of animals and the natural world,[23] I am also centrally concerned with the complementary process of how cross-species encounters are routinely woven into people's *social* worlds and their "urban consciousness." While the child who takes heed of the birds around her might feel more connected to nature, she might also—or might instead—feel more connected to her neighborhood. And glimpsing a rare piping plover is meaningful to the bird-watcher not only because she experiences a novel animal encounter, but also because she shares the experience with a community of birders.

To borrow, in altered form, a phrase from Michael Bell, I am interested in the *social experience of animals*.[24] I heed the insights of the anthropologist Claude Lévi-Strauss and the sociologist Emile Durkheim, who long ago warned against the Nature Lost trope's facile exoticization of the ways that "primitive" people related to nonhumans. Lévi-Strauss refuted the anthropological doctrine, which he called the *totemic illusion*, that clans worshipped nature and viewed animals as literal kin. He saw totemism as "no more than a particular expression, by means of a special nomenclature formed of animal and plant names," of sets of ideas and relations that are "a universal feature of human thinking."[25] Animals were central to totemic belief systems not primarily because they embodied nature but because they were useful symbols for expressing the relationship between self and society. Durkheim reached a similar conclusion in *The Elementary Forms of Religious Life*, arguing that each clan's sacralization of a particular animal was actually a way of sanctifying the collective. Thus, while scholars like Gibson and Berger claim that "authentic" relationships with animals are those that somehow eclipse social life and foster primordial ties to nature, Lévi-Strauss and Durkheim imply that social categories have *always* structured our relations with animals and that—contrary to the biophilia thesis—the sanctity of such relations can sometimes result from their unique capacity to enrich *social* life.

While most social scientists now recognize that "primitive" people are more like "us" than outmoded anthropology texts indicate, few question the assumption that—as sociologist Bruno Latour notes—technology has liberated "modern" society from nature to such an extent that it can now be understood purely through reference to social facts.[26] Certainly, urbanization fosters lifestyles less *directly* dependent on intercourse with nature. But the thesis of a steadily increasing distance between people and other species is too crude. Urbanization has, for instance, brought humans into *greater* contact with a host of synanthropes (e.g., pigeons, crows, rats) that "flourish with the milder climates, abundant food, and protection from harassment that our cities provide."[27] And novel urban initiatives are fostering more cross-species encounters in cities like New York, where new "green spaces" and wildlife corridors are being created, "dog parks" are becoming de rigueur, and a demand for affordable and locally grown food is leading to the proliferation of community gardens, rooftop beehives, and backyard chicken coops.[28]

Conceiving of nonhumans as separate from society—the corollary of Nature Lost—precludes a sociological understanding of how they are incorporated into contemporary social life. Animals *become* pets or pests through social processes of interaction and classification. Studying these processes can reveal how people define themselves through relationships to other species.[29] While sociologists recognize that our sense of self is created through interaction, prevailing accounts of the "social self" neglect the role that nonhumans play in its construction.[30]

Analyzing the social experience of animals does not mean reducing them to passive, symbolic objects. As I will show, animals play an active—if inadvertent—role in shaping how humans interact with and interpret them. Moreover, they are not simply props in the stories people "tell themselves about themselves,"[31] as anthropologist Clifford Geertz would have it—animals *shape* society's stories. Indeed, geographer Chris Philo claims that part of humans' ambivalence about encounters with nonhumans is their capacity to disrupt the social order by "squeez[ing] out of the places, or out of the roles . . . which human beings envisage for them."[32] Certainly, people may rejoice when beautiful or rare untamed flora and fauna grace their yards or ledges. The red-tailed hawk called Pale Male became a Manhattan icon when he built a nest on the ledge of a housing cooperative, using antipigeon spikes as an anchor, and raised a family. We find "wildness" appealing, or at least tolerable, to a point. But past that point, wild becomes socially repulsive (unless, of course, it is "out there" in the wilderness): dogs must be made to obey, excess deer must be culled, and lawns must be mowed. Geographer Yi-Fu Tuan, pointing to the popularity of ornamental gardens and pets, observes that we most enjoy having nature in our midst when we can exercise our "impulse to reduce—and thereby, order and control" it.[33]

One warrant for looking at animals in cities is because these seem to be sites where the ambivalence surrounding cross-species encounters is most salient. After all, most urban spaces are imagined as human-only places—

"nature" is in the "countryside." In our mental maps of the city, animals and plants are relegated to compartmental realms where encounters with them can be managed: the manicured park, the dog run, the zoo, or the tree encased in concrete. But in everyday life, nonhumans overflow these physical and symbolic containers and collide with the sidewalk's pulsing humanity: weeds poke through asphalt; gulls scavenge from dumpsters; feral cats prowl alleys; and diseases like West Nile virus jump from animals to humans. The edges of the city and nature continually rub against, and run over, each other like tectonic plates. The interaction may be smooth, or tremors may result; and the fault line can widen, narrow, and shift as an effect of the encounter. By straddling the fault line, we can begin to understand how the borders and contours of urban experience are shaped *through* cross-species encounters.

By exploring how relations with animals are part of "urbanism as a way of life," this book also engages with classic concerns of urban sociology. For over a century, scholars of city life have sought to understand how strangers from diverse socioeconomic backgrounds manage social relations in dense and anonymous cities, and why social order persists or breaks down. A primary objective has been to delineate the conditions that foster group life, or at least civility, among people who do not necessarily share a history or the ascriptive characteristics that have traditionally bound together villages and rural communities—such as kin and ethnicity.[34] In the pages ahead, I touch on similar questions. I document, for instance, how animals can disrupt the social order of the city—such as when pigeons became Trafalgar Square's public enemy number one—and how relations with animals can foster social ties that transcend ethnicity—such as when Italians, Puerto Ricans, and blacks formed a primary group based on their shared interest in pigeon flying. I hope that this book's case studies speak across the chasms that too often separate scholarship on cities, the environment, and animals.

The Global Pigeon

Because this book does not center on one place or social group, its trajectory is a bit unusual. I "search for odd connections rather than seamless generalizations,"[35] developing several overlapping yet distinct conceptual themes by pairing up the different case studies.

Part 1, "The Pedestrian Pigeon," looks at how street pigeons were knitted into people's experience and imagination of urban public spaces.

I first analyze how routine pigeon-feeding episodes in Father Demo Square enabled visitors to forge meaningful connections to sidewalk life by providing them with easy and noncommittal opportunities for sociable play. I then consider how pigeons make places what they are—and vice versa—through an analysis of two famous sites: in Piazza San Marco, I show how the centuries-old custom of buying seed from vendors and allowing the hungry pigeons to eat from one's hand reified the birds as an embodiment of the square's cultural heritage and elevated feeding to a ritual; in Trafalgar Square, I show how the mayor's controversial eviction of a longtime seed vendor and criminalization of pigeon feeding was an attempt to signal that the square's renovation had erased its reputation as a neglected and disorderly space. These chapters reveal how animals can become part of what sociologist Erving Goffman called the "interaction order,"[36] of public space, and how social contexts structure whether or not people welcome the presence of the "wild" in city streets. They trouble the assumption that animals have disappeared from, or are inconsequential in, everyday urban life; and they show how animals can enchant city sidewalks not by offering a bond with nature but by enabling a felt connection to place and culture.

Part 2, "The Totemic Pigeon," focuses on how relationships with domestic pigeons configured people's sense of self and relation to society. I show how the New York flyers' appreciation for pigeons was given impetus by their social relations: they bred and trained their birds according to group customs, and they competed for peer status vis-à-vis their birds. I also examine how the men's animal practices were steeped in working-class culture and fostered social ties that crossed racial boundaries. Shifting to the Turkish pigeon caretakers in Berlin, I show how the birds were vehicles for the men's performance of their ethnic identity, and I describe how the pigeon coop enabled these homesick immigrants to simultaneously maintain a material connection to their homeland and create a Turkish social space in the interstices of their host city. By underscoring how people's close relations with animals can be driven by socially patterned impulses and can organize the "social self," these chapters critique the notion that such associations are inherently tied to an innate desire to commune with nature. They also challenge sociological perspectives that assume nonhumans play no part in shaping the social realm.

Part 3, "Deep Play," examines the symbolic significance of pigeon racing and traces how the professionalization of the sport is undermining its status as "the poor man's horse racing." While pigeon racing offered financial and egoistic rewards, fanciers' attraction to this animal competition cannot be satisfactorily explained in instrumental terms. For the blue-collar Bronx Club members, I document how pigeon racing was a portal to a magical world in which fanciers were the playthings of nature's transcendent power, and how—win or lose—their birds' return was thrilling because it signified a triumph of nurture over nature and confirmed the value of the labor the men devoted to race preparation. I also underline that competition was meaningful because it rooted fanciers in their urban community. I then show that, beyond the cash prize, the Million Dollar Pigeon Race (MDPR) was seductive because it embodied the ideals of a "flat world" in which competitors from around the globe had equal opportunity to win because all of their birds were subject to the same conditions.[37] The trade-off, though, was a radical alteration of fanciers' relationship to their sport, their community, and their birds: a stratum of paid specialists in Sun City took over the tasks customarily handled by a single fancier, the race was decoupled from local clubs in order to attract a global clientele, and all of the pigeons returned to a single loft at the casino while their owners watched from afar via a video screen. Comparing the MDPR and the Bronx Club reveals how professionalization is deskilling fancying and tipping the scales in favor of moneyed fanciers. Although global races like the MDPR may bring greater prestige and rewards to the sport, their popularity seems to be doing little to halt the decline of local pigeon clubs in urban communities. Thus, while some fanciers interpret the rationalization of their sport as an allegory for the promise of "free and fair" global competition, others seem to see in it the "disenchantment of the world" that haunted Max Weber.

The conclusion, "Changing Ecologies," situates the decline of fancying and growing antipathy toward pigeons within a larger narrative about urban environmental change. I explore tensions between efforts to expunge nonhumans from city sidewalks and efforts to "bring nature back in," and I critique "the myth of asocial nature" as well as sociologists' willful neglect of nonhumans. I propose that urbanism today is marked by "hybridity," and that "seeing the social" in nature can advance efforts to foster greater respect and appreciation for other species.

"The global pigeon" has three distinct meanings. Most basically, it refers to the fact that my research spans three continents and that pigeons—

both feral and domesticated—are everywhere. I could have chosen to look at street pigeons in the public squares of Mexico City or Hong Kong, and I could have chosen to examine the social worlds of pigeon fanciers in Bali or Damascus. This points to the second sense in which I use the term "global." Though I cannot generalize my ethnographic findings to a broader population, I believe that my field sites uncover social processes that can help illuminate other settings. While we would need to observe the sidewalks of Paris to find out *how* visitors interpret the wildlife they encounter there, for instance, I strongly suspect we would still find that nonhumans color pedestrians' experience of urban space (for better or worse). And there is evidence that all sorts of human-animal relations, from pets to livestock, can shape people's "social self."[38] Finally, this book reaches for the global by attempting to uncover how extralocal processes—social, political, and economic—impinged on these particular sites and people, molding the patterns of behavior that I witnessed. In this pursuit, I use each case study as a sensitizing device, holding it up to the others. This helps bring the context in each setting, which is the scaffolding that supports interaction, into sharper relief.

My method of "following the pigeon" is akin to the multisited ethnographic approach that anthropologist George Marcus calls "following the thing." The idea, he says, is to trace "the circulation through different contexts of a manifestly material object of study." Anthropologist Arjun Appadurai points out that simply acknowledging *theoretically* that objects—or, in this case, animals—have situational meanings does little to advance our understanding: "we have to follow the things themselves, for their meanings are inscribed in their forms, their uses, their trajectories . . . it is things-in-motion that illuminate their human and social context."[39]

Appadurai's words remind the ethnographer not to be seduced by theoretical abstractions and tidy typologies. In the pages that follow, I privilege *interaction* and consider the meanings people give to their own actions to be an important part of explaining social behavior. While I find actors' own narratives to be compelling sources of data, one benefit of "being there" is that I can also juxtapose what actors say and what they do in situ. This entails providing detailed accounts of how social life unfolds—the lived experience of the actors involved, the temporal progression of their encounters, and the larger webs of meaning they draw upon. Thus, I attempt to limit my use of summary accounts of interactions and to favor the presentation of specific events and encounters from the field. This approach also allows the reader to interpret the data and assess for herself whether my analyses seem valid.

Throughout this book, 'single quotes' represent my reconstruction of interactions and conversations based on field notes; "double quotes" are the product of recorded and transcribed utterances. Except for a few instances in which my participants requested anonymity, people—and places—are identified using their real names (some opted to withhold surnames). As anthropologist Linda Seligmann notes, this journalistic convention recognizes the value of my participants' lives as individuals and allows others "to contest or enrich this initial research."[40] I use past tense throughout the book to remind the reader that all of my descriptions reflect a historically situated time and place. One can never step in the same river twice.

I

The Pedestrian Pigeon

ONE

Feeding the Pigeons

Sidewalk Sociability in Greenwich Village

FATHER DEMO SQUARE sits in the heart of Greenwich Village, a triangular brick plot hemmed in by Carmine Street, Bleecker Street, and Sixth Avenue. The plaza, created as a leftover when Sixth Avenue was expanded south in 1925, is named in honor of an Italian pastor who raised money to build the church located just west of the park. As part of a study on interactions in public space, I documented the social life of this park—along with five others—from the spring of 2004 until the fall of 2005. The interactions and setting depicted below were all recorded before extensive renovations, which rendered Father Demo Square inaccessible from 2006 to 2007.

Though only 0.072 acres, the plaza's location in a busy commercial and pedestrian crossroads means that it nourishes the kind of vibrant sidewalk life that urbanist Jane Jacobs, who lived in and wrote about the area, claimed was crucial for fostering a sense of social connectedness in the city.[1] Visitors to the park included many who grabbed a slice from Joe's Pizza or a bagel from Bagels on the Square, delivery-truck drivers and city workers, employees of nearby businesses, tourists poring over their maps, students from New York University, club goers and pub crawlers, and the homeless, among others. The most frequent users of the park during the day were the old, mostly white, lifelong Villagers. Some told me that the few hours they spent there each day served as their primary form of social interaction.

The presence of many park visitors who ate takeout food there, as well as people who specifically came to the park to deposit birdseed and bread, meant that Father Demo Square was also a place that attracted pigeons. Despite the "Do Not Feed the Pigeons" sign, visitors provided

the birds with meals all day long. When I spoke with the head of a block association whose jurisdiction included the park, he complained about such visitors: "There are some people who come and leave bags of food for these pigeons, and they don't even live in the community." He also complained about homeless people, who he said often slept in the space and could be found passed out drunk in the morning, but he had hopes that the renovated park, which would be fenced and locked at night, would control the homeless population. The block captain was having a harder time trying to figure out how to curb the pigeon population. In a view echoed by the Parks Department and the local councilperson's office, he called the pigeons a nuisance and a health hazard. He explained the pigeons' presence as the product of a few zealous pigeon-lovers that ruined it for everybody else: "It's only a few people that feed the pigeons; 99.999% of people want to see no pigeons in the park. Everybody complains about it."

My observations of everyday park life painted a different picture. I saw that pigeon feeding was a daily routine for at least a half dozen people who came with seed for the birds, and over twice as many who brought bread. Dozens more park users engaged in more casual and spontaneous pigeon feeding every day, preferring to toss their pizza crust or bagel remains to the birds instead of in the trash. At many moments of the day, pigeons dominated the scene, and many people in the park were engaged with them in some form or another—if not feeding them, then perhaps chasing them (common among children) or photographing them. In fact, human interactions with pigeons seemed to be among the most common "face-to-face" encounters that took place in Father Demo Square.

I was particularly struck by the capacity of pigeons to recognize regular feeders and coax unwitting park visitors into feeding them, behaviors that reflected the birds' adaptation to the demands and opportunities of city living. Biologists have found that pigeons are "able to learn quickly from their interactions with human feeders" on the street and "use this knowledge to maximize the profitability of the urban environment," discriminating between friendly feeders and hostile pedestrians and adopting begging strategies that elicit food from strangers.[2]

While it is obvious what pigeons gain from this synanthropic relationship, this chapter examines the social significance of pigeon feeding. Though the block captain and some municipal authorities framed pigeons as nuisances that impeded people's enjoyment of sidewalk life, I found that pigeon-feeding routines could become part of what Jane Jacobs called the "intricate sidewalk ballet" that enriched pedestrians' experience of the street and satisfied their desire for casual and delimited forms of copresence.[3] It was not uncommon for park visitors to strike up a conversation with a stranger at one moment and then become absorbed in tossing pieces of their meal to pigeons at the next moment. While most sociologists would ignore the human-animal encounter or view it as incommensurable with the social encounter, I saw how both forms of sidewalk interaction offered the solitary pedestrian the simple pleasure of playful, noncommittal association—what classical sociologist Georg Simmel called *sociability*. Pigeons could also instigate spontaneous associations among human strangers by becoming a topic of conversation and a focus of attention. Such interactions reveal how animals can become part of the urban "interaction order," shaping routine sidewalk life.[4] These cross-species encounters also trouble the social psychological axiom that sustained interaction is possible only if participants share a definition of the situation.

Pigeon Feeding

On a mild, sunny November afternoon, I arrived in Father Demo Square around 2:00 p.m. Thirty-two people sat on the benches: some young and some old, some tourists and some locals, some working class in appearance and some in suits. An old white man shuffled through the park, dropping bread as he went but continuing out of the space while looking back to watch 30–35 pigeons swarm the food. This seemed to inspire a Hispanic woman and her daughter, who tossed chips at the pigeons gathered nearby. A middle-aged, disheveled black male shifted his gaze toward about 20 cooing pigeons that gathered expectantly around him. He then tossed some pieces of a bagel to the pigeons and intently watched them consume it, while a young white couple laughed as they watched the birds fight over the food. Two middle-aged white women tossed pieces of bread from their sandwiches to a few pigeons and nonchalantly watched the birds as they continued their conversation. Another older, seemingly homeless black man rummaged through all of the garbage cans and then consumed what he found while breaking off pieces for a dozen pigeons. Two young Hispanic men watched the frenzied activity of the pigeons and laughed, and then began to toss some of their meal to the birds. A second old white man walked through the park and dropped some bread on his way out, and an old white woman on a bench decided to join in the feeding by tossing some bread from her sandwich.

Then Anna, an old white woman who fed the pigeons every day and whom regulars called the "pigeon lady," took a bag of birdseed out of her purse and dumped some of it on the ground. Well over 50 pigeons flew over to her from surrounding rooftops, cooing loudly and shoving each other as they competed for the food. Anna continued to toss the seed about her, and held some in her hands, which encouraged some of the birds to climb on her body to obtain it. She smiled as a pigeon sat on her knee and another on her shoulder. A white woman stopped to take photos of the spectacle and chatted with Anna. At that point, passersby had stopped to observe, laugh, and take pictures of the event. A black boy, likely in middle school, got up from his bench and raced through the middle of the park, laughing and chasing the birds until they took off in a bunch and circled the plaza. Their simultaneous taking of flight let off loud claps of over 100 wings, and spectators pointed and ducked as the pigeons flew barely over their heads.

I was sitting with a young white woman, Carey, and two old white men, Frank and Jerry, who were regulars at the park. Carey remarked, 'It looks like they get fed all day.' They laughed as we watched the pigeons try to carry off large pieces of bagels while others attempted to rip it out of their beaks. Frank commented that the pigeons are here to stay because of all the food, and we focused our attention on them as the birds fought it out. In this not unusual five-minute snapshot from an afternoon in the park, out of the 32 people present inside the boundaries of the space, 18 people were directly involved with the pigeons through feeding, chasing, or taking photographs. More people, both those sitting and passersby, were indirectly involved in the action through taking time out from their other activities to watch the interactions.

Though feeding pigeons in New York (and elsewhere) can result in a nuisance citation, it was a routine activity for many visitors to Father Demo Square and other parks I observed around the city. Feeding pigeons also seemed common in the public spaces of many other cities I have visited, from Notre Dame in Paris to the Plaza de Mayo in Buenos Aires. People engage in pigeon feeding with various levels of intensity. Some people feed pigeons while paying little attention to them, or feed them for short durations. Sociologist Erving Goffman would call these "unfocused interactions." But other people achieve "focused interactions"[5] while feeding pigeons, in which they become absorbed in the activity and engage in feeding for an extended period of time. Though there is no absolute boundary between the two, "focused feeding" looks different than "unfocused feeding," and it seems that these two interactive forms tend to function differently as well. Unfocused feeding often helps facilitate associations *among people*, in which pigeons become a resource for conversations and interactions among acquaintances and even strangers. But focused feeding facilitates human associations *with pigeons*. In focused feeding, humans appear to seek these associations as ends in themselves. As a result of sustained attention to the pigeons, humans are rewarded with increasingly complex interactions. In both forms, pigeons—if only for the moment—help satisfy pedestrians' desire for copresence.

Unfocused feeding, in which humans paid little or sporadic attention to the pigeons, often occurred when two or more people were together. On one occasion, a middle-aged black woman and her middle-aged white female companion entered the park with sandwiches and sat down, while a nearby white family of four was feeding about twenty pigeons. After the women had been eating and talking for about two minutes, occasionally looking in the direction of the family, six pigeons slowly approached to within about eight feet of the women, stretching out and cocking their heads sideways. They seemed to expect food. The black woman appeared to notice them first. She ripped several pieces of bread from her sandwich while she returned eye contact to her friend. She then tossed some crumbs one at a time. One pigeon walked ahead of the others, craning its neck just enough to quickly snatch the crumb while maintaining a distance of about three feet from the women. The black woman noticed this from the corner of her eye, and two of the other pigeons advanced and pecked at the crumbs that were closer to the women. Soon, all six pigeons had moved within four feet of the women, and the black woman ripped off and tossed more pieces of bread, appearing to nonchalantly watch them eat with quick glances while she focused most of her eye contact on her friend. However, her friend had noticed the pigeons and also began to tear off and toss portions of her sandwich bread at them.

The women continued their conversation, laughing and discussing a variety of topics such as work and family life. Yet now both women took regular glances over at the pigeons they were feeding. Their actions brought over 5 more pigeons, and a mass of about 30–35 birds began to slowly make their way toward the women and the 11 pigeons already eating there. At this point, the women trained their eyes on the 11 pigeons longer, and they ceased talking to each other but smiled and laughed as they jointly tossed more and more crumbs. They talked about the pigeons to each other, pointing out ones that appeared quite eager or somehow comical in their actions (e.g., cooing loudly). Yet when the larger group of pigeons came within six feet of them, the white woman, and then the black woman, ceased tossing crumbs.

Only when about 25–30 pigeons had left to seek food from other visitors did the women begin the routine again, going back and forth between eye contact with each other and glances at the birds, making small talk that was sprinkled with references to the pigeons, and taking time-outs to focus their eyes on the pigeons. Six or seven minutes had passed, and the pigeons were now coming quite close to the women (less than one foot) to get some food. Yet, when the white woman moved her foot, the pigeons took a short flight away. The birds appeared to stay cautious despite their desire for more food. Again, they returned slowly. From here on out the women maintained an on and off feeding schedule that alternately attracted and then pushed back the pigeons. The women appeared to work to keep a small number of the pigeons interested, and seemed to enjoy their presence, but they also worked to prevent their association from being overwhelmed by the 40–50 pigeons that were currently in the park.

After about 15 minutes, the women finished their sandwiches and soon after exited the park.

It appeared that the mere presence of the birds encouraged some people to feed them. It was not quite as simple, however, as copresence. Pigeons worked to bring about this result by "begging." If on statues or lampposts, they swooped to the ground when food appeared. They were often on the ground already, because pigeons, like chickens, are naturally inclined to scavenge for food by pecking the ground—not flowers, trees, and shrubs (which is why they walk rather than hop). Pigeons would move in close, within a range of one to six feet, to the person(s) with food. Moreover, they stretched their necks out and turned their heads sideways so that one of their eyes was clearly directed at the food supply. If people ignored the birds, they often approached even closer, sometimes within inches of one's feet. If no food was forthcoming, they might move away, but often they did so only to double back and try again. In the end, they were remarkably adept at getting the average park visitor to donate scraps from their meal. Once the food had disappeared, the pigeons backed away.

Feeding episodes often seemed unplanned, and they were also "contagious." There were often lulls where no one was feeding the pigeons, and then as soon as one started, others joined in. Such feeding usually occurred among families and groups. In fact, the women in the episode described above had already observed a family feeding some pigeons before the black woman began to toss her own crumbs. The white family of four, eating pizza, was tossing pieces of crust to the pigeons. The father began the activity, and as a group of about 20 pigeons gathered around, he verbally encouraged his young son and daughter to toss food. He helped them break off the crumbs and throw them, while saying, 'Look at the birdies. Feed the birdies.' The mother showed her approval, smiling and then grabbing some crust herself to toss to the birds. The children began to laugh, looking back to their parents who reinforced for them (through smiles and continual feeding) that the family as a whole framed this activity as an enjoyable event. All four family members tossed crust and laughed, with the children seemingly delighted to draw the pigeons into their company. The birds became the main involvement of the family, channeling their attention into a communal activity.

Children of varying ages could often be seen stopping by on their way home from school to feed the birds. Groups of adults, including delivery-truck drivers and sanitation workers who took a lunch break there, also incorporated feeding into their routines. Of course, some people

fed pigeons and some did not; some who did it focused more attention on the birds than others; and some talked about the pigeons while others did not. The features that were more or less common to these episodes of unfocused feeding were that they were usually spontaneous, tended to occur with two or more associated people, were short in duration (under two minutes) or sporadic, and resulted in a predictable and narrow range of actions on the part of the pigeons. The birds usually stayed several feet away and simply grabbed the food and ran if the feeder did not invest a longer period of uninterrupted time in them (though pigeons habituated quickly, a sudden move could send them off). I seldom saw a large mass of pigeons around casual feeders. Interestingly, the vast majority of unfocused feeding episodes appeared to be instigated by pigeons, which successfully employed begging strategies until they were rewarded.

A key component of unfocused feeding is that pigeons, while brought into interaction with humans, were often incorporated into sociable exchanges that were still largely carried out among people. Significantly, pigeons became resources that facilitated or enhanced associations among humans in this space. There were only brief moments in unfocused feeding that humans directed their full attention to the pigeons. Thus, the presence of pigeons in the park could be a *means* when they acted as a prop employed to instigate interaction among humans. Of course, in the process humans became a means for pigeons in the birds' everyday pursuit of subsistence.

In public places, strangers are often expected to—at most—briefly acknowledge one another and then divert their attention elsewhere. Although strangers may wish to engage in sidewalk interactions, rules of civility dictate that they need an excuse to do so. Erving Goffman observed that dogs are a "classic bridging device" between strangers in public, and studies of urban parks confirm that dogs "facilitate encounters among the previously unacquainted."[6] In public places like Father Demo Square, pigeons too may act as a sort of interactional prop among strangers—in addition to focusing the attention of those already associated.

As I sat in the park one day, Anna the "pigeon lady" was there, engaged in her routine. A white female tourist who had been wandering by doubled back to witness the pigeons crawling on Anna. After sitting close by and taking some photos, the young woman struck up a conversation with Anna, and they talked for well over 15 minutes. This happened regularly. Indeed, by feeding the pigeons Anna not only obtained their company but also gained the company of strangers she would not otherwise likely interact with. On many occasions, strangers on nearby benches made remarks to one another or struck up sustained conversations based on the pigeon-feeding episodes. Even if people were complaining about the birds, this complaint was often directed to a nearby stranger who was then expected to reciprocate with a response. Such exchanges might last seconds, but occasionally they lasted for over 20 minutes. Once the subject matter of the pigeons had been exhausted, some strangers had worked up enough trust and interest in one another to launch into other subject matters. Other people, however, got drawn into the interactions with the pigeons themselves, seeking these interactions as the ultimate ends of their park visits.

Focused feeding, in which humans engaged in sustained "face-to-face" interaction with pigeons, usually occurred when pigeon feeders were alone, though there were many episodes of its occurrence with pairs of people as well. However, three or more people tended to create an atmosphere in which the pigeons' presence was secondary. One afternoon, I watched an elderly white couple enter the park, the woman pushing the man in a wheelchair. They sat for about three minutes, speaking occasionally to each other as they ate some sandwiches. The woman casually tossed a crumb to a nearby pigeon. This pigeon quickly ate the piece, and the woman then fixed her eyes on it. As about 20 pigeons moved into their proximity, the woman tossed more crumbs to the birds as she smiled and remarked to her companion in Russian. Then he followed her cue, and they continually tossed crumbs as more and more birds gathered around and they maintained their conversation. After about three minutes, over 50 pigeons had crowded around them, cocking and twisting their heads. The couple pointed at the birds and quickened their pace of crumb tossing. In the next minute or so, they entirely ceased their conversation with each other and silently fixed their gazes on the birds gathering about their feet. This continued for about six or seven minutes. The man noticed that some birds approached the base of his wheelchair, so he simply let some crumbs drop right out of his hands. He smiled, and continually monitored the pigeons in his immediate vicinity to decide where to toss his bread next. The pigeons came closer after a couple minutes of constant feeding, willing to go between the legs of the woman to snatch up some crumbs, and she laughed out loud as a few began to jump on the bench and look at her in apparent expectation. The couple appeared completely absorbed, seemingly "lost" in the encounter. It was only when they had run out of food (after about 15 minutes), and the pigeons had moved on to the next feeder, that the couple looked at and spoke to one another. Their smiles faded. Then they looked around quickly, possibly embarrassed, and the woman immediately stood up and wheeled the man out of the park.

I saw this couple feeding the pigeons again three days later, and this time they had purposely brought bread in a bag for the birds. In their previous encounter, they spontaneously and at first casually fed the birds some crumbs, but wound up engaging in a focused interaction. They now returned with a dedicated cache of bread for the pigeons, to ensure a sustained encounter, and they were on their way to becoming regular feeders. Food was an important marker that demonstrated the level of commitment to feeding. Unfocused feeding was not only spontaneous and short lived, but also usually entailed simply breaking off very minor portions of one's bagel, pizza crust, or bread. Many who engaged in focused feeding reserved food just for pigeons, either a separate supply or a generous portion of their own meal. I saw that several apparently homeless men, mostly black, would partake in feeding this way. These men came to the park alone to get food, and after sorting through the garbage cans, they often took a seat and began to eat. It was quite common in this process for some of them to set aside some of the food attained—whether potato chips, noodles, or salad—for the pigeons, tossed on the ground next to the man as he ate, so that it looked as if the pigeons and the man were taking a meal together.

One chilly day I spotted a white male, disheveled and likely in his mid-40s, rummaging through two garbage cans. He found part of a sandwich and some potato chips. He dumped the bag of chips on the ground and then smashed them so that the pigeons could eat them more readily. Then he sat and ate in silence for several minutes, watching the birds. When one came closer to him, he said to it, 'Hey little guy,' and then he broke off a piece of his sandwich for it. While the man did not smile, he appeared enthused, tossing more bread to the gathering pigeons that now stood in a semicircle around him. He talked to them a bit more: 'Here you go'; 'Don't fight, there's plenty for everyone.' Four pigeons alighted on the bench, and one ate only a few inches from his hand. The man experimented, tossing food closer and closer to him to see how near the pigeons would come for food. When he tried to touch one and it jumped back, he returned to feeding them from a "safe" distance of about six inches. When he finished, after about ten minutes, I introduced myself to him. His name was John, and he told me that he slept by a New York University dorm. John said that he felt sorry for the pigeons and could see how hungry they were, but that he also enjoyed their company and fed them to keep them around.

Many parks have their dedicated pigeon feeders, who come nearly every day with bread or seed especially for the pigeons. While a lot of people who feed pigeons shuttle their attention back and forth between pigeons and other humans, or come to the park and then "discover" the pigeons, these individuals come with the primary intention of feeding the birds. Anna, the elderly white woman mentioned above, was one such person—called the "pigeon lady" by seven or eight regulars of this park. Others abound: in Washington Square there was a middle-aged Hispanic man named Teddy. Near Union Square was an older white woman named Mary. When I met Anna, she was living alone in a nearby walk-up apartment. She complained that all of her old friends from the neighborhood had died or been "put in a home." One time I pointed out some of the people whom I had identified as regulars and asked her about them. Anna seemed to know only one woman, who lived in her building. With her arthritis, getting out of her apartment was a chore. Yet Anna showed up to interact with the pigeons three times a day. While Anna was happy to engage in conversation with humans, she came for the pigeons. It is clear that there was a form of regularity and familiarity in the daily interactions that occurred between the pigeons and Anna, one that I did not see between pigeons and casual feeders.

On an early spring day, Anna walked gingerly into the park with her cane in one hand and bags of food in the other. As usual—for her, but a spectacle to many—the pigeons flew down from the rooftops and walked behind her cooing, following her into the park. Anna carefully lowered herself onto a bench. Anna was 85 years old at the time, and claimed that she had been feeding the pigeons for 31 years. On this day I asked her how she started feeding pigeons, and she replied that everyone asked her that but she really did not know. She said that she just decided to begin feeding them one day and had since stayed with it. Anna regularly fed the birds seed and chunks of pita bread that she got for free from a Middle Eastern restaurant.

While I had seen pigeons in the park habituate to feeders over time, getting closer and even willing to jump on the bench with humans or stand on their feet, the pigeons displayed an elevated level of "trust" and familiarity with Anna brought about by her regular routines of prolonged feeding. When she got within a block of the park, they followed her. When she neared the bench, some even alighted on her shoulders. When she sat down, they were already beginning to climb on her; and when she fed them, they ate out of her hand. Anna was able to discern certain pigeons that were familiar to her from a pack in which almost all of them looked the same to me. As one pigeon jumped onto her knee, Anna asked me if I had met her friend Spots. She claimed "he" is a good bird, for when he gets enough to eat he leaves. Anna told me that she had three or four "favorites," Spots and Nasty among them. She wondered aloud where Nasty was, expecting him to be here. I asked her if the pigeons have different personalities, and she told me they do. Elaborating, she said that Nasty picks fights with everyone, and that he always tries to get into her pockets. Just then she noticed that Nasty had arrived. The bird flew down and jumped onto her knee to eat out of a plastic bag of seed that she held in her hand, but, sure enough, against her wishes Nasty began to peck in her pockets.

Anna claimed that the birds expected her to come with food, and she as well expected certain pigeons to be there and to interact with her in certain ways. For example, Anna wondered aloud where Nasty was, and she would get more upset when Spots went into her pocket for food than

when Nasty did, because Spots was supposed to be the "good" pigeon. Anna saw something in the pigeons that many, including me, missed. She could tell the pigeons apart so that she knew that the same birds were showing up every day, and she would get upset if some went missing. Yet Anna's association with the pigeons went further than this. Based on an established routine, a *relationship* emerged. Anna's feeding habits fostered a unique history with the birds that opened up more interactive possibilities with the pigeons than we have seen with the other feeding episodes. The pigeons *did* recognize Anna from across the street and lined up to wait for her, and Anna could point out certain pigeons and predict with accuracy how they would act and which ones would be willing to sit on her lap. Anna and the pigeons mutually constructed coordinated routines, in which their shared history enabled rather complex and distinctive interactive forms.

Anna's sessions with these birds often lasted for over half an hour; and, like the homeless man John, even if she might have been partially motivated to feed the pigeons so that they did not starve, this motivation seemed at most to serve as a baseline that brought them together. Out of the possibly purposive action of feeding sprung engrossing interactions with far more elaborated forms than those that tended to occur with unfocused feeding. Anna could have simply dumped the food and left the park if she did not seek the association as an end in itself, as some people did (perhaps to avoid a citation). Yet Anna "inefficiently" fed the birds by limiting the amount of food given at one time in order to prolong and expand the parameters of the encounter.

Regular pigeon feeders told me that they found satisfaction in having the birds eat from their hand, and in becoming familiar enough with a group of pigeons that they could discern "regulars." These people did more things with pigeons than casual feeders did, forming associations with greater interactive depth and breadth. Focused feeding almost always happened when feeders were alone. This more easily allowed the feeder to become attuned to the birds' responses, to lose herself in the "flow" of reciprocal interaction. While focused feeding was still common with pairs, any number more than two associated people tended to lead to sporadic and relatively unabsorbed involvement with the pigeons. It seems that the presence of more people threatened to "flood out" the membrane of the human-pigeon encounter by opening up the possibility of distracting "side engrossments" among people.[7] Yet for some people, interacting with pigeons seemed to be an end in itself that provided them with the satisfaction of association, even if it was a more limited form of association than interaction with other humans.

Focused feeding was usually longer in duration (ten or more minutes) and more continuous (uninterrupted) than casual feeding. This fostered more variability in the routines of interaction. Pigeons sometimes emerged as individual personalities, some people talked to them, the pigeons landed on shoulders and ate out of hands, and so forth. Humans experimented with different actions to see what they could "get away with" and still keep the pigeons involved in the encounter. Pigeons and humans were in sync with the actions of the other(s), based on prior history and prolonged interactions. One example is that, when Anna lifted her leg, the pigeons did not scurry away the way they did when casual feeders made a similar sudden move. The essential element of focused feeding was that pigeons ceased to be bit players in the background of street scenes; they became the center of attention for humans, who became absorbed in these "focused interactions." To be sure, other humans might still be part of the picture—but often were not. Also, even if interaction with other humans did occur, it was usually sporadic and appeared to be mostly just a slight addition to the interaction context, in the same way that the pigeons often were merely an addition to the interaction context of humans in unfocused feeding.

The experiential difference between focused and unfocused feeding is significant. The pigeons were there, in the park, an open possibility for anyone to encounter and interact with in a variety of ways. Some ignored them, and some fed them but just as a minor diversion or as a resource for facilitating their socializing with friends or strangers. But in focused feeding, it appears that individuals became engrossed, even "lost," in the associative possibilities of these encounters. Goffman astutely remarked that "solitary playfulness will give way to sociable playfulness when a usable other appears, which, in many cases, can be a member of another species."[8] Pigeons, not just people, can serve as "usable others" for those seeking sidewalk sociability. But how is this interaction coordinated across such a stark human-animal divide?[9]

Asymmetrical Interaction

According to the social psychological paradigm of "symbolic interactionism," sustained interaction depends on participants' ability to mutually negotiate meanings and imaginatively "take the role of the other" so that they can create a shared definition of the situation.[10] This excludes human-animal encounters from the "interaction order" because the absence of language makes symmetrical understandings impossible. Human-animal researchers, however, have demonstrated the depth and complexity of people's everyday interactions with their pet dogs and cats, in which they jointly construct elaborate routines that are marked by bodily communication and attunement to each other's actions. A few scholars have taken this as evidence that humans and their pets *do* share symmetrical understandings of the situation and have some ability to take the role of the other. Clinton Sanders and Leslie Irvine, for example, argue that games like fetch are possible only because both human and dog can read and send signals that symbolize "this is play."[11] But must humans and animals share symbols to interact? This assumption seems unverifiable. Moreover, while such a theory may help explain how people interact with dogs, it seems that few, if any, would argue that humans and pigeons interact through shared symbols.

An alternative way to understand cross-species interaction is to bracket the collective act and analyze each participant's particular behaviors and strategies. In studying human-dog play, psychologists Robert Mitchell and Nicholas Thompson found that dogs and humans had their own "projects" that they tried to achieve through interaction.[12] For example, a human might have the project of keeping the ball away from the dog, whose project might be to acquire the ball. As long as the flow of the encounters provided both human and dog with sufficient opportunities to pursue their own project, sustained interaction was possible—even if humans were not psychologically engaged in such encounters (e.g., were bored) or were simply trying to give the dog some exercise. This indicates that *compatible projects* are sufficient to produce coordinated interaction, whether shared meanings and goals exist or not. Interacting parties, then, can have different intentions in "successful" interactions and can assign *asymmetrical meanings* to them.[13] This is certainly the case with pigeon feeding (and likely so with adult-infant interactions).

The behaviors of street pigeons indicated that they had a persistent goal of obtaining food, and they sought out situations in which they could obtain it. This was how humans found pigeons in the park, lurking about and waiting for someone to toss some food on the ground. Further, the pigeons attempted to manipulate the actions of park visitors to ensure the desired result by "begging." People in the park could have any number of projects, but some of them encountered the pigeons in this setting and began feeding them. This may be an example of two purposive projects—pigeons want to eat, and maybe humans want to help nourish these birds. In focused feeding, though, the person centers her attention on the activity. Often, she seeks to elaborate on the interactive possibilities and thus has a more durable encounter. Her project has ceased to be about "just" feeding the birds. Out of this purposive project have arisen additional actions aimed at obtaining playful association and reciprocal interactions with these creatures.

Though the encounters with pigeons followed one basic structure—feeding—humans often performed this task with intentionally "inefficient" moves (e.g., giving one crumb at a time) in order to sustain the flow of the encounter.[14] Interaction took on an ordered, routinized form. Yet people could use the feeding routine as a foundation on which to elaborate and improvise, much as sociable talk among park visitors might consist of "the elaboration of a conversational resource."[15] In focused feeding, humans appeared to become lost in the flow of the encounter; they allowed themselves to be enveloped in these momentary associations. Its reward was its own repetition. And because pigeons—like most animals—were capable of action independent of direct human manipulation, they could also add an enjoyable spice of uncertainty for humans.

Though we may know that pigeons in the park want to eat when they approach us, we can totally disregard their mental states and simply enjoy feeding them. However, "successful" pigeon feeding relies on some amount of monitoring of pigeon habits (e.g., if you walk after pigeons while tossing food, they will run away). Dedicated feeders were able to obtain enhanced satisfaction in the association itself, and their repeated experiences over time resulted in an increased mutual attunement between humans and pigeons that allowed for more elaborated forms of interaction (e.g., the pigeons were no longer afraid when the feeder came near, and sometimes they were quick to eat from her hand or stand on her knee; one could distinguish pigeon "personalities"). There is, then, work involved in achieving engrossing interactions with pigeons. Compared to, say, dogs, there are fewer possible varieties of interaction. *In form*, however, focused pigeon feeding—like Anna's interactions—still reflects a mutually constructed, relatively complex association.

Feeding pigeons has the potential to be engrossing and meaningful to humans, regardless of how the interaction is experienced by pigeons, because the "projects" or intentions of the two species are compatible with each other—at this moment, in this setting—even as those projects have entirely different aims. However different its intentions, the pigeon still confronts the feeder as an active agent, what philosopher Martin Buber would call a "thou," "a truly *subjective* other whose immediate presence is compelling."[16] The pigeon can initiate and reciprocate action in a number of ways that are neither predetermined nor entirely under the control of people. These traits are present in almost all animals. Yet many of them are not encountered in a city dweller's everyday life, nor do their "projects" play out in ways that would keep them close enough to people to sustain the "face-to-face" interaction possible among humans and pigeons. For example, raccoons in city parks are less likely to open themselves up to such encounters.

Long ago, Georg Simmel observed that even in the midst of instrumental social interactions, there is often an "impulse" to enhance them, to create moments of sheer satisfaction by association.[17] Such sociable play can grow out of the topsoil of purposive action. In the same way, we see that even if one feeds pigeons for the purpose of providing nourishment, from here may grow an "impulse" to elaborate on this here-and-now interaction by introducing interactive forms for no other purpose than to enhance the association. It is this "impulse" to associate as an end in itself, and the fact that the pigeon brings something out of the person that enables both of them to accomplish structures of interaction together, that permits people to experience pigeon feeding as a compelling and fulfilling form of sidewalk sociability.

Sidewalk Sociability and Social Vulnerability

Following Jane Jacobs, urban sociologists have found that noncommittal sidewalk sociability, and even people watching, can mitigate feelings of isolation and foster a sense of connection to the urban social fabric.[18] However, while public spaces provide opportunities for spontaneous interactions, these interactions are not always coveted. Indeed, Jacobs observed that many pedestrians seek to balance a desire for copresence with a demand for privacy. And Goffman noted that, on sidewalks, people from all walks of life are "exposed" to the possibility of being engaged in unwanted interactions with other individuals who, granted the public space they share, have the "right" to initiate encounters that may not be granted in other contexts.[19] An example of this, which I saw many times in Father Demo Square, is when a seemingly homeless person intentionally sits close to a park visitor and initiates a conversation. The visitor, "caught" and not wanting to appear rude, must interact with the homeless person for some duration before it is acceptable to walk away. There is a subtle tension at work, then, when one enters "open regions" such as parks. By putting oneself in a public place, the individual is open to the desirable possibility of engaging in sociable encounters with strangers *and* open to unwanted social entanglements or even social isolation.[20] We are made vulnerable.

Interactions with pigeons took place in this context. Especially if people came to the park alone—as most people who engaged in focused feeding did—they were "open," or "exposed." People often visited the park to take a "time-out" from their instrumental concerns. As Goffman remarked, to avoid appearing strange or conspicuous, unless one engages in people watching, a sort of prop is needed to give the appearance of a "main involvement." Indeed, my observations reveal that people watching may serve as a "legitimate" activity only for a limited duration. Understanding the park visitor in this way, it becomes easy to see how some visitors wound up feeding pigeons. On a basic level, it was something to do that made one appear occupied.

Further, I have almost never seen a pigeon reject food. While this may seem trivial, the fact that it takes such an easily obtainable prop (food) to interact with a pigeon in a way that the person largely controls, and the knowledge beforehand that the pigeons will immediately gather around the person (and maybe even climb on her if one is patient) in the hopes of obtaining more food, assures the individual that it will be quite easy to achieve an association with these creatures. The matter is much less trivial for people who may spend many hours each day in social isolation or experience a sense of marginalization from the larger community, such as some elderly and homeless people. The image of the old, friendless "pigeon lady," an urban archetype reinforced in stories and movies such as *Mary Poppins* and *Home Alone 2*, resonates for a reason. Though I found regular pigeon feeders to be more diverse than the stereotype—including married men, the wealthy, and the middle-aged—they did tend to be older people living alone. Similarly, biologists who studied pigeon feeding in Madrid and Basel described regular feeders as "the lonely, old, disabled—those not accepted in general."[21] Though at first I was surprised to see so many homeless people feeding pigeons, I came to see how the easily secured associations that these episodes afforded could be a powerful lure for those who were generally shunned by people on the sidewalks. Encounters with pigeons ephemerally dissolved their solitude into an experience of "togetherness" with others. In the process, feeding provided a narrative structure to idle time, as humans instigated behaviors from pigeons (e.g., begging) which in turn required a human response to complete the sequence. Pigeons needed them.

We should shrink from generalizations, though. Perhaps those most eager to feed pigeons were children, who seemed to care less about obeying park rules that forbade the act and who did not yet seem to harbor notions of pigeons as pests. My observations also reveal that casual feeders far outnumbered regular feeders—and those who spontaneously fed pigeons came from all walks of life. The thing that most casual feeders did share was the *social situation of finding themselves alone, and idle, in the park*. In other words, for the moment they experienced the more generalized condition of some of the socially isolated regular feeders, and they responded in kind. Thus, the "impulse" to feed pigeons, which ecologists and nature writers like Edward O. Wilson might construe as the expression of an "innate" psychological desire to commune with other species,[22] may not be so distinct from the impulse to begin chatting with a stranger: they may both be born out of a situational response to solitude in a moment of unstructured time.

The difference between casual and regular feeders was one of degree rather than kind. As I have shown, pigeons could coax unsuspecting park visitors into feeding them. Unfocused feeding readily bled into focused feeding, particularly if a person was alone. And satisfying feeding episodes could beget future encounters. It is easy to see how casual feeders can become regulars. The "career" of a pigeon feeder is not so mysterious or deviant, and her activities and aims may not be so different from other park visitors'.

Pigeons offer possibilities for interaction that newspapers or people watching do not, as Anna's story reveals. Being with pigeons means not being alone. Feeding possesses a "consummatory end-in-itself character,"[23] lending structure and meaning to free time. Outside concerns can be momentarily dissolved; the individual is free to become "lost" in the activity. The satisfaction can precisely be found in investing one's full attention and consciousness into an association in which so little is at stake. Akin to one type of social contact that Jacobs said typified "successful" sidewalk life, pigeons can offer enjoyable, casual associations that penetrate the pedestrian's blasé and reserved public persona while entailing little commitment.[24]

Pigeons are a visible, active part of the street scene. While they may be irrelevant or annoying to some people, to others pigeons provide a ready resource for interaction. They may serve as a prop for associations among people, but they may also become the focus of park visitors' attention. Pigeons play a part in determining this outcome by initiating feeding episodes and sustaining interactions in their quest for food. By virtue of being enmeshed in these relations with people, they can play a prominent role in shaping how people interact in, and experience, urban spaces. The complex interactive routines that regular feeders forge with pigeons challenge the social psychological assumption that coordinated face-to-face encounters require shared symbols. And the sociability that people can achieve through pigeon feeding—especially for solitary pedestrians—demonstrates that mundane encounters with animals can offer opportunities for the kind of informal, bounded forms of copresence that urban sociologists claim enrich people's experience of the sidewalk and combat feelings of social isolation.

Everyday feeding episodes like the ones that I witnessed in Greenwich Village also have material consequences. Pigeons frequent locations where they are fed; if they are fed well, they breed profusely; and because they are rewarded with food for begging, they decrease the time they spend scavenging for food around the city and become more dependent on people. The large number and vast spatial distribution of pigeons, and the problems that such populations may engender, are the direct product of human activity. This socioecological reality indicates how popular pigeon feeding still is, despite its growing stigma.

TWO

"Do Not Feed the Pigeons"

Cultural Heritage and the Politics of Place in Venice and London

PIAZZA SAN MARCO and Trafalgar Square are among the most well known public spaces in the world, and pigeons have been legendary denizens of both. It seems, in fact, that their historic association with flocks of feral pigeons is a defining feature of these spaces. The block captain in Greenwich Village understood this. Lamenting the frequency of pigeon feeding in Father Demo Square, he complained that the space is "not Piazza San Marco." He did not want pigeons to become a celebrated feature of his local park, as they clearly were in Piazza San Marco, and he admired London's bold mayor, who put a halt to the pigeon-feeding tradition in Trafalgar Square. I realized that these squares were "theoretically strategic research sites" for examining how understandings of urban spaces are constructed in relation to pigeons, because they manifested, with exaggerated clarity, the contrasting possibilities that I saw in more "diluted" form in New York.[1] I spent the month of June 2006 observing and participating in the street life of these squares, and I also performed nineteen interviews and perused four municipal archives.

In the stately Piazza San Marco, for two weeks I joined throngs of tourists who bought packets of seed from licensed vendors and then posed like statues as an estimated 20,000 pigeons swarmed them throughout the daylight hours. (Pigeons are diurnal, or active in the day. They tend to retire to their roosts before nightfall.) This spectacle was a centuries-old ritual of the piazza. In London, instead of the many pigeons that once called Trafalgar Square home, I found a man patrolling the space with a trained Harris's hawk on his arm—a literal manifestation of the mayor's hawkish stance on pigeons. After the mayor announced plans to "clean up" the square, he evicted its seed vendor and criminalized pigeon feeding. Yet, upon incurring the wrath of concerned Londoners who argued that the tame pigeons would starve, in subsequent years a bizarre arrangement—which I observed—held in the square: each morning, a group of volunteers would dump over 100 pounds of birdseed supplied by the city onto the ground to make up for the feed no longer supplied by the vendor.

    I found it striking that in Venice, pigeons were part of the very essence of the piazza as a unique place, prominently featured in postcards, paintings, and guidebooks. Even municipal leaders I spoke with who lamented the corrosion of monuments that the birds hastened said that they could not imagine the space without its feathered icons. In London, however, the mayor promoted a new vision for Trafalgar Square, a cultural monument that had to be reclaimed from its status as a traffic island and a haven for what he called rats with wings. His denigration of pigeons as filthy—even taboo—objects, as well as his feeding ban and his hiring of a falconer[2] to repel the birds, were central to his efforts to exorcise the old, unsavory image of the square.

    While *space* refers to an area's physical properties, *place* refers to its social meanings. Places embody history. Sociologist Michael Bell writes that the social history of a place can be thought of as a *ghost* because it animates—or haunts—our lived experience of location, shaping our interpretations and interactions. We also endow spaces with our own meanings, forming "a social tie with the physical world" as a result of personally significant experiences in a location that enable us "to sense our own ghost in [the] place."[3] In Piazza San Marco and Trafalgar Square, it was apparent that people's interactions with pigeons were a primary way by which these spaces became meaningful and knowable to them. It was *through* feeding pigeons that people experienced the piazza as a place, forming unique memories. But visitors' experiences were patterned, and their molds were found ready-made in the form of congealed place-based customs. The piazza normalized and even compelled pigeon feeding by framing the pigeons as cultural attractions. In Trafalgar Square, however, "Do Not Feed the Pigeons" signs, the presence of park wardens who enforced the rule, and the menacing scowl of a predatory hawk presented a very different message. The place defined pigeon feeding as deviant, and London's mayor framed the birds as antithetical to the makeover of Trafalgar Square as a "cultural space." However, many Londoners sought to channel the ghost of place by continuing to feed the birds.

    These two squares highlight how understandings of animals are tethered to understandings of space and place. While the ubiquity of pigeons and other urban synanthropes contradicts the well-worn premise that nonhumans have disappeared from everyday city life, London's mayor mobilized the imaginary *ideal* of urban spaces as human-only places in order to redefine pigeons as "out of place" in Trafalgar Square. His opponents, along with visitors to Piazza San Marco, instead interpreted the pigeons as legitimate and even fêted sidewalk habitués because the custom of pigeon feeding was a constitutive component of these squares' cultural heritage. The political conflict over pigeons in these two squares—Venice's mayor controversially decided in 2008 to join Trafalgar Square in banning the sale of pigeon feed—illustrates how interpretations of animals are patterned both by enduring cultural templates that frame the presence of the "wild" in city spaces as disorder and by local *contexts* that can contradict this spatial logic by incorporating nonhumans into the very definition of urban places.

The Pigeons of Piazza San Marco, Venice

The city of Venice is laid over 117 tiny islands, crisscrossed by 150 canals and 409 footbridges. Water taxis and the legendary bowed gondolas take the place of cars. Upon disappearing into the serpentine alleys, one could almost imagine herself in medieval times—except for the 12–14 million annual tourists who clog these nooks and render invisible the local population of 64,000.

Stumbling upon Piazza San Marco for the first time startled me. This massive space—almost the size of a football field, flooded with light, and offering views of the bay—starkly contrasted with the dark maze of the surrounding alleys. The piazza is flanked on three sides by palaces, now museums or offices, with arcades containing shops and cafés on the ground floor. The Byzantine-inspired Basilica di San Marco dominates the east side. A red brick bell tower (the Campanile) stands on the diagonal from the basilica. Around the corner from the Campanile is the Piazzetta di San Marco, which extends from the piazza to the water like the short leg of an *L* and hosts the Gothic "doge's palace" (Palazzo Ducale). Two columns topped by statues of San Teodoro and the Lion of San Marco frame views of gondoliers and the southern bay.

The legend of how an estimated 20,000 pigeons came to call the piazza home is apocryphal. It is said that, in centuries past, the doge (chief magistrate) released pigeons from the gallery of the basilica on Palm Sunday.

Most were caught and eaten, considered an Easter gift to the hungry populace. But the "rugged survivors were felt to have earned St. Mark's protection" and were thus granted "permanent immunity."[4] They reproduced and prospered ever since, as pigeon feeding took on sacred undertones. In an 1851 letter by the English poet Elizabeth Barrett Browning, she described how her son "had made friends with the 'holy pigeons,' and they were surrounding him like a cloud today for the sake of his piece of bread . . . he stamping and crying out for rapture in the grand piazza."[5]

Photographers began to profit from such encounters. A pre–World War II Baedeker guidebook observed, "Those whose ambitions lean in that direction . . . may have themselves photographed covered with the birds."[6] For decades, the pigeons were fed twice daily by the city, in addition to being fed by visitors. An insurance company took over this feeding job from the city in the 1950s as a means of advertisement. The city also issued licenses to sell seed as part of a broader granting of licenses to merchandise and souvenir vendors.

Because of pigeons' storied history in the piazza, they have been called the "most celebrated of the Venetian fauna" and "the living creature that the world associates instinctively with Venice."[7] The historian and travel writer Jan Morris claimed, "Sometimes invalid pigeons . . . become known individually to the waiters at Piazza cafes, and are thereafter privileged for life, allowed to preen themselves on unoccupied tables, and fed wonderfully sustaining morsels of toasted sandwich . . . Pigeons can get away with almost anything in Venice."[8]

While iconic, the pigeons have also been problematic. Since the 1960s, city officials have included pigeons in a constellation of problems facing Venice: pollution, the sinking of the city, unsustainable levels of tourists, deindustrialization, and decaying buildings. Most complaints cite the damage pigeons' acidic feces cause to monuments, and the diseases they may carry. The city employed nets, spikes, birth control–laced feed, and even predatory crows to try to control pigeons.[9] Although a 1999 survey found that 80% of Venetians thought the city should reduce the number of pigeons, curbing pigeon feeding in the piazza remained an unpopular option.[10] A travel essay articulated why: "Though the pigeon . . . is sometimes looked upon as a creature non grata by urban authorities, in Venice . . . the local citizenry views him as a special kind of religious symbol and good luck charm" of the piazza. The article noted that the piazza's pigeons have "posed for more photos than any other group of animals in history" and are "one of the great unexpected attractions of Northern Italy."[11]

Paradoxically, pigeons were simultaneously framed as part of the city's filth that needed to be cleaned up to promote tourism *and* as a signature tourist attraction. Even as Venice's mayor announced a plan in 1999 to cull thousands of pigeons as part of an effort to make over Venice's image, a "Venetian Hotel" opened in Las Vegas, complete with a replica of the doge's palace and a Piazza San Marco featuring live pigeons. The mayor lost his bid for reelection in 2000, leaving the pigeons and seed vendors untouched when I arrived in late May of 2006.

Despite the stunning setting of what Napoleon called "the finest drawing room in Europe," when I first entered the piazza, my attention was overwhelmed by the activities unfolding in it. While hundreds of small outdoor-café tables lined two sides of the square, the rest of the empty space was filled by hundreds of standing and kneeling tourists offering food to thousands of pigeons. Vendors who occupied small mobile stands sold 3.5-ounce packets filled with dried corn and peas for €1. Professional photographers jockeyed to immortalize the scenes for tourists, who usually had their own cameras. Hordes of visitors continually spilled out of the alleys, took a moment to watch feeders in action, and then marched over to the vendors to partake in the practice themselves.

It was hectic. There were perhaps 1,500 people in the piazza along with thousands of audacious pigeons, who scrambled between feet to capture every morsel and descended by the dozens onto tourists' arms and shoulders before the feed packets were even opened. Their heads pumped like jackhammers on people's palms. Groups of tourists returned repeatedly to the vendors for more seed, readying their cameras to capture the encounter. I had never seen pigeons like this. Due to a history of feeding, they "trusted" any human that had food. They even alighted on my shoulder as I took out a notepad, anticipating food. I also noticed some pigeons pecking at leftovers on empty tables, or preening themselves on the backs of chairs, at the legendary Caffè Lavena as a tuxedoed band played to patrons sipping €10 espressos.

I decided to give it a try, though I was slightly hesitant. The sight of dozens of these ravenous pigeons flying straight at me was enough to invoke Hitchcock's dystopian film *The Birds*. I wondered if I would get defecated on, and if the pigeons' scaly feet and claws, or their hard beaks, would cut my skin. Yet my timidity seemed absurd, for around me were

hundreds of people feeding the birds with abandon and letting them crawl all over their bodies.

As I poured the morsels into my cupped hand, one by one pigeons plunked down on my arms and wrists. They pecked with precision, leaving me unharmed, but they were frantic, standing on top of and beating one another with their wings to secure more food and flapping continuously to keep balance. I placed some feed on my shoulders and head after watching others do the same. A half-dozen birds unabashedly landed on my shoulders, and two or three stood on my uncovered head. I felt satisfied that I could coax them to land where I wanted them. My laughing joined a chorus from other tourists as we knowingly smiled and commented on the novelty of the situation. I counted over 40 people feeding pigeons within just 50 feet of me. Our collective behavior set the scene for newly arriving tourists, who observed and then joined us as this ritual was reproduced into the evening.

In the two weeks I was there, I normally saw 7–9 seed vendors at a time in the piazza and piazzetta between 9:00 a.m. and 8:30 p.m. They worked six-hour shifts. There were usually 3–5 photographers present, as well as 12–15 stands selling cheap souvenirs. Workers swept up the discarded seed packets throughout the day. The most intense feeding periods were before and after lunch, when I counted three packets sold per minute by one vendor. But location was key, and vendors were required to rotate daily among standard spots. Near the basilica was best. Farther away, even during a busy time a vendor might sell only one bag every two minutes.

I befriended a vendor named Wesley, a gentle man in his 40s with a blonde ponytail and a mustache, who had emigrated from Poland. Like several other vendors, Wesley did not own his cart. Rather, the owner, who inherited the license, provided the cart and the seed and hired Wesley on commission. For every packet sold, Wesley got 40 euro-cents and the owner got 60 cents. His wages, which he tracked on his cell phone, were unpredictable. One day he made only €35; days later, he made over €150.

CAPTURING AND REPRODUCING THE VENICE EXPERIENCE

Near-identical pigeon-feeding episodes repeated for hours each day. It looked to me like a mass choreographed performance. A key factor in producing this scripted response was the way that the image of Piazza San Marco extended beyond its physical confines through the airwaves, Internet, and printed page. Few people likely came to the piazza totally unprepared for what they would encounter. Guidebooks, travel magazines, and tour guides described the role of pigeons in the piazza and told visitors where to obtain feed. I saw this representation being reproduced. A film crew spent one morning shooting scenes for the James Bond remake of *Casino Royale*. As they meticulously set the stage for a moment in which Bond would run through the piazza, workers dispersed packets of seed right before the director yelled "action." When Bond sprinted through the masses of tourists, the gaggles of scattering pigeons cemented the scene's authenticity. Just 50 feet away, a crew was simultaneously filming a travel segment for *The Today Show*. A blond English woman in her mid-30s stood as men sprinkled pigeon seed in her hands moments before they began recording. As the birds perched on her arms, she spoke into the camera: 'I am in the very famous St. Mark's Square, the absolute heart of Venice and everyone's first port of call in this beautiful city. Hopefully you like birds!'

The extralocal linkage of pigeon feeding and the piazza structured visitors' expectations and helped shape acquiescent participants.[12] And, whether visitors entered through the airport or train station, they were greeted by postcards and paintings portraying pigeon feeding over the years. I also saw black-and-white posters all around town from a time long ago when Coca-Cola spelled out its logo in pigeons on the piazza. Such

representations situated pigeon feeding as a timeless piazza ritual. In the square, powerful cues further guided visitors toward feeding. By hovering around newcomers even before they purchased seed, the tame birds displayed their expectations that people would feed them and in fact "trained" people to do so. Visitors also found identical vending stands, affixed with licenses, providing grain in prepackaged envelopes. And photographers' stands displayed photos of smiling tourists covered in pigeons.

The most powerful feature that guided newcomers toward feeding was the presence of hundreds of exuberant pigeon feeders of almost every age and ethnicity. They not only lured newcomers to feed, they provided models for how to do so. Newcomers often stood back and watched others feed the birds. Sometimes one bold member of a timid group would step forth as the "guinea pig." All would watch as he or she enticed the birds, and then they would buy their own feed to ensure that they experienced this unique phenomenon. As I watched one group of teens, I heard quotes such as 'This is awesome! This is the best time I've had in Venice! I can't believe there's pigeons on me—take a picture!' One girl screamed triumphantly, 'I did it! I was so scared, but I did it!' When the first boy to feed the birds lay on the ground as the pigeons swarmed him, for the moment he became the hero of his cheering peers. Those who had at first appeared too timid to feed the birds were emboldened by such scenes.

Tourists all seemed to want the same thing—to make photos and videos with pigeons all over them—and verbally instructed each other on how to get it: stand still, cup your hands, extend your arms straight from your sides. Thus, visitors spontaneously socialized newcomers, abetting chain reactions. People kept purchasing feed until they captured the scenes they wanted, even when it meant that a just-married couple got pigeon feces on their wedding attire.

Pigeon feeding was actually one of the few legitimate piazza activities. A sign warned, "It is forbidden to sit down in spaces not specifically designated for this purpose . . . [or] to stop to eat or drink anywhere other than at the tables set out by public restaurants."[13] Because the only seats around were at the cafés, visitors' two main legal options were to pay €10 to sit for coffee (the cheapest menu item) or to pay €1 to feed the pigeons (one could, theoretically, just stand around). I did, however, see some people sit unmolested on the stairs.

The sign was a tacit articulation of a message the piazza projected explicitly: feeding pigeons is an expected, valid practice. This is important. Of dozens of visitors I spoke with, only two said they had ever fed pigeons before. Worldwide, feeding pigeons is an increasingly illegal, and thus deviant, activity. Yet the piazza *normalized* this potentially stigmatizing behavior. Feeding the birds invoked a "when in Rome" sensibility that reversed taboo. I watched, for instance, a female tour guide dare a male tourist to join her in lying on the ground as a cameraman filmed them, saying 'This will be on your DVD!' Though the man lamented, 'But there's poop all over the ground,' he lay down next to her and allowed the birds to peck at their still bodies like vultures while a crowd roared its approval. People sometimes seemed aware that they were enacting what sociologist John Urry calls the "tourist gaze," taking a break from their normal routines and roles in order to experience the extraordinary.[14] While I observed two young women from Australia, one joked to the other as she danced with the pigeons, 'I'm the pigeon lady!' And I overheard one woman ask another after they had finished feeding the birds, 'Do you feel different about pigeons?' She responded, 'No, I still think they're disgusting.'

The remarkable consistency of people's response to pigeons was a *social property of the piazza*, which clearly played a large hand in producing these interactions, not merely serving as a backdrop for them.[15] A bracketed space was created in which people became open to interacting with pigeons in novel ways, evidenced in a recurring situation in which people initially showed disgust at the birds and a tentativeness to engage them but then eventually fed them and enjoyed it. As I watched eight young American women of varying ethnicities prepare for their encounter, one shouted, 'Oh my God! I *hate* pigeons!' Her friend recoiled and responded, 'Me too! I don't want to hold any!' As some of their friends fed the birds, they scrunched up their faces in apparent disgust. But eventually, all the women were smiling and appeared at ease while pigeons clung to their arms. I watched a similar scenario play out nearby with an 11–12-year-old French boy. As his mother tried to snap a photo of him with the pigeons, he threw his hands up and cried. Yet the parents were unrelenting, buying packet after packet as they tried to bring their son around to feeding the birds. It worked. He eventually smiled when a bird alighted on his father's hand. After 20 minutes of gradual exposure, he allowed the birds to land on him while his parents dutifully took pictures and he flashed a huge grin.

The pigeons themselves were routinely interpreted as distinctive, as when one young white man claimed that the 'pigeons we have back home suck' compared to the piazza's pigeons, or when a middle-aged black man who talked to the pigeons eating from his hand turned to his wife and remarked, 'They ain't nothing like the ones in New York!' On another occasion, a white mother from Texas offered me hand sanitizer and asked, 'Do you feed the pigeons at home?' When I responded no, she countered, 'Me neither. There's something special about Italian pigeons.' Of this appeal of the piazza's birds, two cultural critics point out disdainfully, "It is interesting to notice, in the midst of this most remarkably ornate and historically rich of ambients, how much enthusiasm and energy is put into attracting, observing, and photographing such a mundane, ordinary, and rather dirty bird."[16] But these jaded observers miss the point. The mythical connotations of pigeon feeding in the square, and the ways that the "socialized" birds provided tourists with spontaneous and novel interactions, endowed the Piazza San Marco birds with a *magical* quality. Akin to what sociologist Jonathan Wynn calls "urban alchemy," these ordinary birds became extraordinary, enchanting people's experience of the piazza.[17]

While there is merit to the cultural critique that tourists were on a quixotic quest for the "authentic" Venice and that feeding pigeons was a hackneyed way to consume that fantasy,[18] to each actor pigeon feeding was *lived* as unique. Feeding the birds enabled an interactive engagement with the piazza and was a way for visitors to personally experience and consecrate the square as a distinct place. Unlike the "focused feeding" I observed in Father Demo Square, it did not seem that people fed pigeons

in the piazza as a form of sidewalk sociability—almost no one fed the birds alone, and tourists' incessant desire to document the interactions and to perform ever more wild antics for the camera seemed aimed at creating singular memories of their time there. Nor did pigeons enchant the square because they offered people a bridge to nature. The pigeons enchanted the square because through them visitors could forge an intimate connection with what Michael Bell calls the "aura" or "spirit" of the place. Although our "imaginative geography" routinely leads us to define *rural* places through their connection to animals, in the piazza pigeons were foundational to people's conception of a thoroughly *urban* space.[19]

MODERN PROBLEMS AND THE WEIGHT OF HISTORY

While pigeons had long been celebrated in Piazza San Marco, they seemed to be a growing headache for the municipality. Massimo Cacciari, who planned to drastically reduce their numbers before losing the mayoral office in 2000, was reelected in 2005. Determined to "project a cleaner image" of his city, he renewed efforts to bring pigeon feeding in the piazza to a halt.[20]

This tension between the celebratory frame on the square and the problematic frame of the government was evident in my conversations (in English) with Venetian municipal workers. At the tourist office, an employee regaled me with the legend of the pigeons' origins before, unprompted, adding that the city saw them as a nuisance that threatened monuments. At the Environmental Office, which served as an informational clearinghouse on Venetian ecology, I spoke with "Michael Bidinger."[21] Michael, a gregarious man in his 40s with dark eyes and salt-and-pepper hair, complained that pigeons left behind 'kilos of shit' in the city's streets and that nothing was being done. What he told me next seemed incomprehensible: since 1997, a city ordinance had prohibited the "direct administration of food to feral pigeons" throughout "the City Council area"—including the piazza—and authorized "an administrative sanction of 1,000,000 lire [€516.45]" for transgressors.[22] When I asked Michael how it could be that the law both legalized selling pigeon feed but criminalized feeding pigeons, he shrugged and said a phrase I would hear from others, 'That's the Italian way.'[23] He added that traditions are sacred, constraining the government from actually ending this historic activity even after it criminalized feeding. Michael's own thinking reflected this ambivalence. After a rant in which he compared seed vendors to the mafia, I asked him what should be done. After a long pause, he sighed and said in a low tone, 'It's difficult to think of San Marco without pigeons. It's a tradition. It's a job.'[24]

Michael helped me arrange an interview with the *assessore*, a powerful city executive appointed by the mayor. Laura Fincato, whose elegant office in City Hall overlooked the Grand Canal, was a professional woman in her 40s with blonde hair, brown eyes, and a strong command of English. After complaining about the city's dwindling population and lamenting that tourism had overrun the city, she said that the filth pigeons created was a 'major problem.' She too called the contradiction of sanctioning licenses for vendors while criminalizing feeding 'the Italian way to escape the problem.' Laura asked me if I had seen the piazza early in the morning—before the pigeons arrived—and sighed, 'Oh, it is so beautiful.' But she immediately followed by saying, 'If you think of the pictures people take in San Marco, you see the pigeons.' She went on to list famous American dignitaries who posed with the birds, such as Bill Clinton. I asked her what she would like to see done. A modest reduction in their numbers seemed to be all she could hope for, and as I got up, she exclaimed, 'It's a tradition.'

Even in their complaints about the damage wrought by pigeons, the city officials I spoke with recognized that the birds were integral to people's sentimental sense of place in the piazza. Though they grumbled that the wheels of bureaucracy seemed to spin in place, they counseled proceeding with caution. Removing the pigeons, even if necessary, would be interpreted as a real symbolic loss.

The Pigeons of Trafalgar Square, London

From Venice, I flew to London to document the struggle over pigeons that the British tabloids called "The Battle of Trafalgar Square." Ever since the Greater London Authority (GLA) rescinded the license of Trafalgar Square's bird-feed vendor in 2001 to make room for "cultural events," pigeon sympathizers had been dumping massive quantities of seed on the square for the birds. The battle had not let up five years later, with the mayor employing hawks to keep the birds away and a group of supporters suing the GLA in the High Court for animal cruelty via starvation. My 15 days in London coincided with the conflict's climax.

I first approached Trafalgar Square after making my way past Buckingham Palace, down "The Mall" boulevard bedecked with Union Jack flags, and through the grand Admiralty Arch. The historian Rodney Mace calls Trafalgar Square the "front room" of England, a place that "attempts to give palpable expression to its . . . social, historical and political aspirations."[25] The square is literally in the center of London, as all distances are officially measured from the junction of major roads at its southern end, known as Charing Cross. The large, open, classical square, built in 1840, commemorates an 1805 sea battle off the coast of Spain in which Vice-Admiral Horatio Nelson led the British navy in a rout of Spanish and French forces. Nelson's Column dominates the square, a behemoth granite pillar jutting over 150 feet in the air and topped by an

18-foot statue. The likeness of Nelson, who died in the battle of Trafalgar, glances south down the artery of Whitehall Road to the Palace of Westminster. Four bronze lions made from the cannons of defeated French ships frame the base of the column, which was considered so central to London's identity that Hitler hoped to abscond to Berlin with it as a war trophy. Three sets of stairs emanating from the square's northern border lead to two large balconies and a permanently closed street abutting the Romanesque, domed National Gallery. Two large fountains and their wide, clover-shaped pools flank the square's eastern and western portions, and recently added bathrooms and a small café hide under the balconies. Yet, with hardly any other ornamentation, some Londoners complain that the square feels stark. Movable signs warn pedestrians in English, French, Arabic, and several other languages, "Do Not Feed the Pigeons. They cause nuisance and damage the square."

Though pigeon feeding in Trafalgar Square does not have an ancient and mythical origin as in Venice, by World War II, the birds, feed vendors, and photographers were ensconced in the space.[26] The foreword of the book *Trafalgar Square through the Camera*, which features a black-and-white cover photo of a laughing woman in a peacoat with pigeons on her head, begins, "One Sunday afternoon when I was ten years old my mother and father decided to take my younger brother and I to Trafalgar Square . . . to be photographed feeding the rather grubby pigeons." Next to a photo of two smiling boys with pigeons in hand dated 1945, Don McCullin writes, "A week later to great joy and astonishment the pictures turned up."[27] Page after page depicts men, women, and children from a bygone era posing in their Sunday best among the birds (and licensed photographers catering to them); and a tender 1940 photo shows a soldier cradling pigeons in his arms. As I combed the Westminster archive's media files, I found piles of news clippings and advertisements from the 1950s onward about the square's pigeons, such as a 1961 ad for Lucky Strike cigarettes from *Time* magazine in which a sharply dressed white male kneels with a cigarette in one hand and a pigeon in the other as his smoking wife, dressed in white, and his blonde daughter look on gleefully. As in Venice, however, pigeons were also sometimes portrayed as a nuisance. A 1967 news story, for instance, warned that the multiplying birds "may rapidly become a serious danger to health and to ancient buildings."[28]

The image of Trafalgar Square suffered in the 1980s, when the vacant and neglected "traffic island" was viewed as "hostile and alienating."[29] In

1989, the writer Charles Owen complained that pedestrian access was difficult, and that there was little seating and little to do. In an impassioned plea to make the square "live up to the eulogies of Baedeker," he envisioned pedestrianizing the roads around the periphery and creating an "enquiries pavilion" as well as a food court with movable tables and chairs. "These facilities being for the people rather than the birds," Owen remarked, pigeon feeding would have to be banned.

The idea to overhaul "London's Place de la Concorde" picked up steam as the city grew concerned about "quality of life" issues that made tourists and residents wary of public spaces. The city council cracked down on petty crimes and illegal street vendors, and the Royal Society of Arts planned to finally bring a sculpture to the square's infamous "fourth plinth," which had remained empty for 150 years. In this context, the city noted the irony of licensing a seed vendor while paying £100,000 annually to clean the square and suggested that the vendor might have to go. However, people reacted in horror when a man was caught taking pigeons from the square in 1996. The magistrate declared, "These pigeons are not any old pigeons but are there to bring character to the area."[30] The tourist board denied the city's claim that the pigeons were a significant health risk and called them "one of London's famous sights."[31] And the late Tony Banks, a Labour Party politician, protested that depriving "these gentle and intelligent creatures" of food would be "unspeakable cruelty."[32] While the pigeon debate was partly framed as public health versus animal cruelty, both sides clearly asserted competing imaginaries of place in which pigeons were either rightful members or trespassers. Pigeon advocates stressed that "the teeming tourist-photographed flocks are as much a colourful and traditional part of the square as . . . Nelson's column," and they wondered how the hometown of the beloved Mary Poppins, who memorably encouraged Londoners to purchase feed from an old "pigeon lady" for "tuppence a bag," could get rid of its "famous" flock.[33] But city officials were forward-looking. The pigeons' alleged filth was tied to the filth and congestion that the city saw as choking the square.

The ambivalent status of Trafalgar Square's pigeons persisted until Ken Livingstone became London's first directly elected mayor in 2000. As head of the GLA, one of his first official acts was to take over the management of Trafalgar Square from Westminster Council.[34] He declared that the pigeons "should all be shot" and that it was time to stop "that bloody pigeon-feed seller. It is the heart of one of the great cities and there is nothing in it except pigeon shit."[35] "I want to see a space," Livingstone declared, "with a programme of events celebrating the variety of arts and cultures . . . [and] creating a communal area that is safe and clean for all people to enjoy."[36] The mayor contended that the "cultural space" he envisioned required that he "maintain good order on the square."[37] While the square was routinely filled with human refuse, the mayor defined pigeons as the major cause of disorder.

Livingstone's ability to vanquish the birds became such a central part of his political agenda that the *Evening Standard* quipped, "Any fool can defeat Tube privatization, hire more policemen and build more affordable homes. But it will take a municipal Napoleon to drive the pigeons from Trafalgar Square . . . Here at last is a test of the mayoral mettle."[38] Bernie Rayner, whose family held a license to vend feed in the square since 1956,[39] sued the GLA—with the support of People for the Ethical Treatment of Animals (PETA)—after his license was rescinded. The mayor had to cancel a ceremonial announcement of his renewal plans when activists threatened to dump pigeon feces on his head, and the above-mentioned member of Parliament Tony Banks joined citizens in staging "feed ins" at the square and charging Livingstone with animal cruelty. The mayor, who was on record as a supporter of animal welfare, was pressured into adopting a "phased withdrawal" in which the GLA provided the birds with a diminishing amount of feed every day for three months, theoretically giving them time to find other food sources. Bernie Rayner dropped his appeal and accepted a buyout for an undisclosed sum. "Everything comes to an end," he lamented, "like the coal works, the steel works."[40] However, a group called the Pigeon Alliance began distributing upward of 220 pounds of feed—paid for with public donations—on the square every day as soon as the phased withdrawal ended, contending that the pigeons did not have enough time to adjust.[41] Forced to cut another deal, the GLA allowed the pigeon advocates, newly organized under the name Save the Trafalgar Square Pigeons (STTSP), to determine the course of the reduction and to distribute feed every morning. In exchange, STTSP signed an agreement not to feed the birds at any other time and to accept the presence of a falconer in the square. Both parties jointly paid independent scientists to monitor the birds' health. An official feeding ban was enacted in 2003.[42]

In remarkable time, Livingstone realized his vision for Trafalgar Square. Traffic was banished from the northern border. A small café and bathrooms were installed, and the perpetually empty fourth plinth was now (controversially) occupied by rotating modern sculptures. A hired falconer patrolled the area, and those who dared feed the birds were fined, including Paris Hilton, who famously protested, "But I love feeding the pigeons in Trafalgar Square . . . I even prefer it to going shopping."[43] Throughout the day, the square appeared pigeon-free. Yet, while most Londoners were still asleep, volunteers were dumping kilos of feed in the square to sustain the flock. This arrangement was supposed to continue until 2008. However, in 2006 Livingstone threatened to void it because of "rogue feeders" who exploited a legal loophole that enabled them to continue feeding pigeons in front of the National Gallery—on the former street that the mayor had pedestrianized—because it was under the jurisdiction of Westminster, not the GLA. This was the state of affairs when I arrived in London in June 2006.

My first stop in London was the visitor's center. While a young man behind the counter said he did not know the history of Trafalgar Square's pigeons, he volunteered, 'They were thinking of putting poison in the feed to kill them but people got upset. I thought it was a great idea.'

I arrived at 1:00 p.m. on a hot and sunny Sunday to find the square busy and festive. A large stage had been erected at the bottom of the central stairs to celebrate Mauritius Day. Under colorful tents and banner ads, vendors sold food, flags, and sports jerseys. A band of black musicians came on stage, playing ska and reggae-inspired music while girls in colorful dresses performed a dance. There were perhaps 300 people watching, though many more revelers milled about. The crowd seemed very diverse: Asians, Arabs, Africans, and whites. People danced, sat on the rims of the fountains, and lay on small patches of grass in front of the National Gallery. The event was over by early evening, and the stage disassembled by about 8:00 p.m. Crowds lingered by the fountains and peripheral benches, and a circle of people practiced capoeira by the National Gallery, drawing a crowd of about 100. The serenity of the square was occasionally punctured by loud, intoxicated young men cruising for women. But by 9:00 the place had settled; perhaps 125 people were scattered throughout the space.

The scene I witnessed was a manifestation of Livingstone's dream. A "cultural event" transformed the space into an outdoor amphitheater that showcased the diversity of the city. "Heritage wardens" prevented people from tossing crumbs for pigeons; people could go to the bathroom onsite; lines formed for the small, industrial-chic café; and people observed the scene from the pedestrianized north terrace. In my time in London, I witnessed several more events like this. The next day was Hare Krishna Day, and the following Saturday was Sri Lanka Day. Yet all of those events paled in comparison to the Bollywood Festival. A massive stage was erected between the fountains, and rows of stadium seating were placed over the stairs. The space was barricaded, with an overflowing crowd inside. These events were partly financed by the companies that displayed large advertisements and sold goods and services at the events.

THE VESTIGES OF THE OLD TRAFALGAR SQUARE

Before my arrival in London I had been in touch with Neil Hanson, the public relations person for STTSP, who invited me to witness their daily 7:30 a.m. feeding. On a dreary Monday morning, I made my way into the quiet and empty square. I saw about 300 pigeons tightly clustered in a corner, cooing loudly and seeming restless. Nearby, four sanitation workers sporting bright green vests swept the square. Occasionally, an apathetic pedestrian or two hurriedly passed through on their way somewhere else. At 7:30 sharp, two white men in their 20s converged on the square, shook hands, and grabbed two large bags of feed (25 kilos, or over 50 pounds, each) from metal storage bins attached to the wall. Perhaps over a thousand more pigeons swooped down from surrounding ledges for a feeding frenzy. Both men dumped the seed uniformly, working their way inward in the shape of a corkscrew as the dizzying mass of gray birds apportioned itself to form a spiral of head-bobbing, flapping, and pecking pigeons.

I introduced myself to the men. Maciej (pronounced "magic") Ruszkowski, an affable young Polish man with blue eyes and combed brown hair who was wearing jeans and a T-shirt, sheepishly admitted that he began feeding the birds 16 months ago because he was unemployed and saw an ad that STTSP was paying £10 per day (about $18) for feeders (most feeders were in fact unpaid). But Maciej grew attached to the flock, looking for his favorites and catching injured birds and taking them to a sanctuary. As he glanced adoringly at the feeding flock, Maciej said in accented English, 'It's incredible to me that they could make it illegal to feed pigeons. OK, I agree they make a mess. But so do people. The square must be cleaned every day regardless.' As we spoke, the sanitation workers scooped water out of the fountains and began to scrub the ground with brushes. Maciej waved at the workers and told me that relations between them were much better than they were before the feeding agreement was reached, when workers allegedly killed pigeons with pressure hoses and vacuums. In just 15 minutes, the birds consumed all of the food and vanished. Maciej and the other man departed soon after.

A half hour later, the falconer Dave Bishop parked his Van Vynck

Avian Solutions van on the square and sipped coffee with the supervisor of the cleaning crew. Dave, whose unique occupation made him a local celebrity, wore a blue-collared work shirt with an embroidered name tag. A white man in his mid-30s, he sported arm tattoos and short, tousled blond hair. Though hesitant to talk with me because the pigeons were such a 'political thing,' Dave seemed proud of his work. 'If you could've seen this place before, you couldn't even see the ground!' As a point of reference, he asked, 'You've seen that bloody Mary Poppins scene?' But now, 'I only see one over there and about seven by that statue.' The hawk, named Emu, did not come out much anymore, as Dave's 'hard work' was done and it was only a matter of maintenance. He laughed, 'In the beginning . . . it was like a drive-through for the hawk. Now, the pigeons are wise to him.' Though the tabloids had a field day when it was revealed that £250,000 of public money had been spent on the hawk (£86 for every estimated pigeon removed), the results were evident.[44]

Aside from occasional tour groups and a lunchtime rush, for most of the day the foot traffic seemed sparse, given the square's size and location, perhaps hosting a few dozen people at a time. But heritage wardens closely policed behavior, stopping a white girl and an Indian man from tossing crumbs to some bold pigeons, kicking people out of the fountains, and warning folks not to sit on the balcony ledge. Dave and the cleaners went home around 5:00.

By witnessing and participating in the feeding sessions over the next two weeks, I saw that these episodes were a time to eulogize, and briefly reconnect with, a place that no longer existed.[45] To the STTSP feeders, a number of whom had been active in other animal-welfare campaigns, the loss of the birds also signaled something tragic about human-animal relations in the city.

One day I watched June, a petite white woman in her late 50s with blonde hair who was renowned by her peers for her tireless work on behalf of the birds, deftly snatch a pigeon and unwind a string from its leg. Then she grabbed another pigeon with a bloated middle toe, held on by only a sliver of skin, and placed it in a canvas bag to take to a sanctuary. She sighed, 'They're such lovely creatures. We kill them, kick them, starve them, and still they trust us.' As she spoke and dispersed seed, pigeons alighted on our shoulders. June went on, 'The Trafalgar Square pigeons are a tradition. For generations people have enjoyed them. Why do we have to feel like convicts for nourishing living creatures?' After we finished feeding the pigeons, we made our way to the bucolic St. James Park so that she could feed the geese and ducks. The next day, Shelagh Moorhouse, a white woman in her 60s with short blonde hair and heavy makeup, nostalgically recalled when it was 'terribly acceptable' to feed the square's thousands of pigeons. While photographing them, she said that her daughters considered feeding the Trafalgar Square pigeons a highlight of their childhood. She complained that people brought more filth to the square than the pigeons. 'I've never seen a pigeon leave a used condom in the fountain.'

When the GLA rolled out grass on the square for a youth soccer match, Shelagh asked, 'Does that make sense when places like Hyde Park and St. James Park are nearby?' She added that the square was ill-suited for the large crowds and temporary stages that typified the mayor's "cultural events." Two of the cleaners (including a supervisor) and two heritage wardens whom I spoke with shared Shelagh's perspective. The burly supervisor grumbled that it was impractical and expensive to set up, break down, and clean up the events and that it was damaging the square. While the wardens and cleaners did not seem to miss the pigeons, they indicated that there was a "political" component to hosting these spectacles on the square. Such events offered proof that the mayor had succeeded in realizing what he called "my vision" for a new Trafalgar Square.

One afternoon I saw hundreds of pigeons descend on the north terrace. I walked over to find a middle-aged woman and a younger man and woman, all white, dispersing seed in front of the Gallery. They fed the pigeons slowly, as if relishing bringing the pedestrian traffic to a standstill. The young woman, Paula, who wore black clothes, purple eyeliner, and a pin that read "Hunt saboteurs save lives," told me that they were from the Pigeon Action Group.

As crowds gawked at the swarming pigeons, Paula told me she was involved in animal rights full time. The older woman, Maxine, was her mother, and she said that all of them were vegetarian and had partaken in protests and campaigns against fox hunting and animal experiments. They made clear that they had nothing to do with STTSP. Maxine bragged, 'They call us the rogue feeders.' Once a part of STTSP, Maxine and some others quit because they refused to sign the agreement to cease feeding in the square. Since then, their activities fanned the GLA's desire to cancel its phased withdrawal program. These self-described radicals said that if Westminster Council made it illegal to feed pigeons on the north terrace, they would resort to "guerilla feeding." The heritage wardens looked on disapprovingly from the stairs of Trafalgar Square but were powerless to stop the feeders. Some passersby asked for seed so that they too could entice pigeons to land on them, and even more took photos.

While both the GLA and STTSP blamed the continued presence of pigeons in front of the Gallery on the "rogue feeders" of the Pigeon Action Group, the reality seemed more complex. While I saw these activists feeding the birds only that one time, it was quite common for visitors, including tourists, to feed pigeons. The history of the square as a haven for pigeons could not be expunged as easily as the vendor Bernie Rayner. Collective memory haunted the space, fostered by diehard feeders, pigeons still tame enough to land on visitors, and the countless images of pigeons in the square still circulating in the media and literature. Though there was considerable variation from day to day, sometimes the north terrace felt like a mini Piazza San Marco.

On one of my first visits to the north terrace, I saw a South Asian man and his children tossing bread at a group of pigeons at around 3:00 p.m. As the children squealed at the bold birds that jumped on their hands,

I saw an eccentric old Indian man approach, wearing crooked sunglasses over his prescription spectacles, a tattered suit jacket, and a blue baseball cap. He pulled some kernels of corn from his pocket and, holding his hand out, got a pigeon to softly touch down on his wrist. At this, the family smiled, and another South Asian family stopped to admire the scene. The Indian man said to one boy, 'Come here, hold out your hand. Hold still. There.' A pigeon jumped on the boy's hand. A small crowd began to gather, and he did the same "trick" with a little girl. The man was putting on a show of sorts, attracting over 20 people as he got the pigeons to land on others' arms, shoulders, and heads by placing corn on them. Passersby were being sucked in, circulating to the man, getting their feed, and reproducing feeding episodes like the ones that used to take place on the square. Birds would even land on the Indian man when he did not have food, and he would "hand off" his pigeons to interested parties and help them 'make the picture,' as he said. 'There used to be thousands of pigeons; they tried to get rid of them,' he told me. 'You can't feed them there anymore,' he said, pointing to the square. 'But you can feed them here.' I asked him why he gave seed to passersby: 'People come from all over the world—they have cameras. The pigeons are stars. They have been in the movies, like *Mary Poppins*. They are famous, so I help people make the picture.'

The old man's routine, while strange and comical, was grounded in the history of the place. He was, in some ways, a standard bearer for the Trafalgar Square that still existed in the minds of many people—a stand-in for Bernie Rayner.[46] While he was hardly the only regular feeder, he seemed zealously driven to ensure that the public continued experiencing the ritual that had defined the square for so long. As soon as he approached the terrace, pigeons recognized him and swooped down. When I stood next to him, even if neither of us had food, they jumped on my shoulder. This gave him the odd appearance of possessing some special power to attract the birds, and he (along with perhaps a half dozen other regulars) routinely instigated feeding sessions that reproduced themselves even after he had left. Just as in Venice, the birds also trained passersby to feed them, including people who did not know that there used to be a seed vendor. I frequently observed locals and return tourists commenting on how there used to be thousands of pigeons and a vendor; and one heritage warden told me that people often asked her where the pigeons were and wondered why it became illegal to feed them.

THE CONTESTED TRAFALGAR SQUARE EXPERIENCE

Livingstone's improvements to the square were impressive. So many people walked between the north terrace and the square that I had trouble imagining that traffic once separated the two. The café was popular, and the bathrooms allowed tourists to linger. The large free events were well attended, and I saw spontaneous gatherings as well, such as spirited celebrations of England's World Cup victories. In decades past, however, Trafalgar Square also served as a venue for what one might call "cultural events." It has long been *the* place to be for New Year's Eve, and the documents I viewed in the Westminster archives depicted big events on the square from decades past such as Maori dancers and a calypso band. One photo of a packed crowd watching the Maori dancers reveals a massive flock of pigeons circling above them around Nelson's Column. At the time, the presence of pigeons on the square was not framed as incongruent with such events.[47]

Trafalgar Square was at a low point in the 1980s, as were many urban spaces: dirty, crumbling, dangerous at night, and host to illicit activities. It was only when this image of Trafalgar Square formed the dominant narrative that pigeons came to symbolize the squalor. Thus, as Livingstone aggressively sought to remake the square, removing the pigeons carried enormous symbolic heft as an indication that the quality of life had been improved. A civilized, orderly, colorful, and safe space meant a space that had zero tolerance for the gray scavengers.

It seemed both infuriating and humiliating to Livingstone that feeding persisted on the north terrace after the Trafalgar Square feeding ban. Eager to free himself from the agreement for the phased withdrawal of food by catching "rogue feeders" or by demonstrating that the birds' numbers were not being reduced (which was the rationale for the agreement), the GLA relied on the wardens, falconers, and even biologists as hired investigative agents.[48]

One morning I noted an older white man with glasses and a clipboard hovering at a distance on the edge of the square. An STTSP feeder named Molga Salvalaggio informed me that he was a biologist sent by the GLA to monitor feeding and the number of birds. When Molga gently asked him what he was doing, he replied briskly, 'I'm trying to count pigeons and you're interrupting me!' As Molga walked away, the biologist seethed to me, 'They're crazy. They say they care about animal welfare, but they don't.' He claimed that the number of birds had barely gone down since 2000. 'It's because the organization still feeds them, despite the agreement.' He grew agitated, saying, 'It's absolute lies to say we're starving the pigeons!' Seemingly recognizing the centrality of claims to place, however, the biologist assured me that the birds had always lived and bred somewhere else. As he glared at Molga, he barked, 'So there is *no such thing* as the Trafalgar Square flock!'

Dave Bishop, one of two falconers who roamed the square, was on the front lines of the mayor's battle against pigeons. This made him the sworn enemy of the STTSP feeders. The feeders claimed that he had made threats against them and that he encouraged "blood sport" on the square. Dave simply said that he had a job to do and that his hawk usually only frightened the birds. Most of the time the square was empty of birds, and Dave left the hawk in the van. But if he decided that pigeons were encroaching on the space, he would place the regal, statuesque bird on his thick leather glove and walk around. Often, just seeing the bird was enough to scare the pigeons off. The hawk was a hit with tourists, who took pictures and asked Dave all sorts of questions about it. I saw Dave release the hawk only once. It glided low and startled a few tourists as it grabbed a pigeon lounging by the stairs in its talons. The pigeon escaped but appeared injured, and the hawk landed on the balcony in front of the Gallery with a clutch of feathers in its claws.

Perhaps visitors to the "new" Trafalgar Square wanted to have their cake and eat it too. In my time in London, it seemed that for many visitors the presence of pigeons was not as antithetical to the new square as it was for the mayor. I saw both children and adults feed birds despite five large signs that announced it was prohibited; heritage wardens constantly shuttled around the square to stop offenders; and wardens stood by helplessly as people simply followed the pigeons and other feeders to the north terrace. Much to Livingstone's chagrin, the road that he pedestrianized became the new locus of the diminished but still vibrant Trafalgar Square flock. Despite all of the explicit cues that discouraged pigeon feeding, it persisted to such a high degree that Livingstone's characterization of pigeon feeders as a small renegade band of "antisocial" activists hardly seemed accurate.

On June 20, however, Livingstone made good on his threat to withdraw from the feeding agreement because of "rogue feeders." The bins of feed were gone. STTSP had been expecting this and initiated proceedings against the GLA in the High Court for breach of contract. The

maintenance supervisor sighed, 'Now the circus begins. There will be reporters. They'll feed the birds all over.' Shelagh, Molga, and Maciej arrived on the north terrace, each carrying large canvas bags of grain. Maciej pointed to the line that separated Westminster from the GLA's jurisdiction, and we began dumping seed. The pigeons soon caught on and swarmed from the square to the terrace. When we returned the next morning, a camera crew from NBC's *The Today Show* was waiting, highlighting how famous the battle over the square's pigeons had become. The aired segment began with stock footage of smiling kids feeding pigeons as a female voiceover said, "For generations, they've been a part of the landscape of London." After cutting to a clip of the mayor shouting to a jeering crowd, "They are rats with wings but without the intelligence," it showed Maciej and Molga walking up to the terrace and dumping large quantities of seed, pointing out the legal loophole. Then it too played up the alleged "rogue feeders," naming the Pigeon Action Group as a "shadowy" organization that was "ruining it all."

Feeding by the Gallery was not ideal. There were far more morning pedestrians than in the square, making the birds vulnerable and likely to incense citizens and Westminster Council. In fact, some pedestrians yelled insults at us such as "Bugger off!" Westminster was already working to secure a bylaw to ban feeding, and STTSP recognized that if they continued to feed on the north terrace they would hasten that result. I walked away on my last day in the square knowing that the situation was unsustainable and sensing that the High Court would rule in the GLA's favor. I left behind a white woman who had just dumped a bag of seed while an Asian family fed bread to the birds. Thirteen people crouched with their hands out to attract the pigeons, an older Indian woman sat on the ground as they ate from her lap, and over 60 people watched. Just as I caught a final glimpse of this scene before crossing the street, I saw the old Indian man appear, no doubt ready to help tourists 'make the picture.'

Before flying home, I made one last stop at Hyde Park Corner to investigate a rumor that the former Trafalgar Square seed vendor worked there. Given the extensive media coverage of his eviction, I easily recognized Bernie Rayner behind the counter of a burger stand. The man denied his identity to me and said 'no comment,' but he betrayed himself with a sad smile and his parting words: 'If Bernie were here, he'd refer you to the GLA. They're the ones with the power.'

Pigeons, Space, and Place

While pigeons did contribute to costly cleanups of Trafalgar Square, many people resisted Livingstone's framing of them as nuisances. Over time, the pigeons had taken on distinctive placed-based meanings that led a lot of Londoners to interpret the birds' eviction as an assault on tradition and culture. The first appeals to prevent the removal of the birds centered on their historical linkage to the square. It was largely after STTSP failed to stop Livingstone's plans on these grounds that it refocused the argument on animal cruelty. Conversely, arguments centered on economics and public health did not do enough to convince the public that the Trafalgar

Square pigeons should be removed; so Livingstone fought fire with fire. He produced a countercultural narrative in which he was banishing the filth and disorder that besmirched this public landmark. The mayor was decisively breaking with history, building a new "cultural space" in which dirty birds were out of place. The remodeled space carried new moral connotations. Pigeon feeding was now taboo. Opposed to old, wholesome images of families and soldiers enjoying the birds, now those who dared feed the birds were labeled "rogue feeders."

While pigeons once enchanted Trafalgar Square and Piazza San Marco, they are now framed as contamination. When Massimo Cacciari returned as the mayor of Venice in 2005, he once again attacked the pigeon problem. Few expected that he would be able to overcome tradition, the 20 vendors locally known as the "pigeon lobby," or the "Italian way" of dealing with problems, which allegedly is to do nothing. But city officials drew strength from Livingstone's example. To the amazement of millions around the world, on April 30, 2008, the city began issuing €50 fines for feeding the birds in Piazza San Marco; it also implemented a ban on selling feed. While the mayor cited evidence of property damage, the vendors denied that pigeon feces were a primary cause of monuments' deterioration and argued that the city was simply trying to take further control of tourism in Venice. "Sooner or later," one vendor warned, "they'll even take away the gondoliers."[49] Thousands of pigeons remain in the square, landing on visitors as they have been habituated to do. And daring tourists bring their own bits of food in order to recreate the classic experience. But the vendors are gone, and so it seems is one of the last bastions of the Western world where pigeons were an institutionally celebrated aspect of place. Like London's mayor, Cacciari was set on projecting a cleaner image of his city, and in the "rat with wings" he found the perfect symbol of the dirt he sought to remove.

By 2007, London's Westminster Council approved a pigeon-feeding ban, closing the legal loophole that had enabled activists and visitors to continue to feed pigeons in front of the National Gallery because it is a stone's throw outside of London proper. While the mayor claimed that his moves "reclaimed [the square] as a pleasant and enjoyable place," it seemed to be giving some Londoners an identity crisis. "It was pleasant and enjoyable with the pigeons," one columnist lamented, "and they made it look like London. How many movies have you seen that featured a panoramic shot of Trafalgar Square with a huge flock of pigeons rising in the air? That was all you needed to tell where the film was set."[50] As an act of resistance against "the loss of sentimental and social connections to [the] place,"⁵¹ as well as out of humane considerations, locals continue to protest at the square, hold candlelight vigils, and dump large amounts of grain for the thousands of tame pigeons that still consider the space home.

A major part of how pigeons were framed as a problem in both squares was by being hitched to the mayors' "quality of life" agenda, which sought to sanitize and bring greater order to public space. This was most salient in the case of London, where Livingstone redeveloped the entire square and explicitly stated that, as a "cultural space," there would be no room for pigeons. The label "rats with wings" appears to have been born out of this linkage to social disorder. In a 1966 *New York Times* article, the New York City parks commissioner Thomas P. Hoving announced a plan to restore Bryant Park. He excoriated litterers and vandals, while the supervisor of Bryant Park lamented, "The homosexuals . . . make faces at people [and] once the winos are dried out at Bellevue, they make a beeline for Bryant Park." The article portrayed a park in crisis, overrun by perceived social ills of the time such as vandalism, litter, the homeless, and homosexuals. Under a heading titled "And There's the Pigeons," the park supervisor called the birds "our most persistent vandal," because "the pigeon eats our ivy, our grass, our flowers, and presents a health menace." Tagged on to the end of this paragraph is the first reported utterance of the metaphor that has stuck to the pigeon ever since: "Commissioner Hoving calls the pigeon 'a rat with wings.'" With that, the article closed with hope that a cleanup would "bring in a better element of people."⁵²

Geographers Chris Philo and Chris Wilbert argue that society imposes "complex spatial expectations"⁵³ upon animals. Dogs that roam free, for example, are considered abandoned or feral and can be impounded—pets "belong" on a leash or in the house or yard. Wolves are deemed pests when they "trespass" on agricultural land, where they prey on farmers' livestock—wild megafauna "belong" in pristine wilds far removed from society, or in a zoo. This "imaginative geography of animals," as Philo and Wilbert label it, suggests that our social and moral evaluations of animals are contingent on where they are found. As a matter of course, humans work to ensure that animals stay in their "proper place." Drawing on the anthropological insights of Mary Douglas, Philo claims that animals that "transgress" the "socio-spatial order" that humans have constructed around them become interpreted as "matter out of place."⁵⁴

If, as William Cronon writes, our ideal conception of nature is pristine wilderness, then our ideal conception of the city is an orderly grid where nonhumans are kept under control and boxed into manicured settings such as a park or a flowerpot. Though this tidy imaginary is an illusion, it persists in the collective consciousness and organizes the ways people experience their environment.[55] The mayors of London and Venice employed this spatial logic to try to extricate the unique local meanings of the squares' birds and frame them as disorder.

Like weeds in the cracks of pavement, pigeons represent chaotic, untamed nature in spaces designated for humans. Pigeons have become particularly despised urban trespassers partly because they, in all their animality, are so *public*. It is almost as if they taunt us with their seemingly "unnatural" predilection for stone and concrete. Conditioned by the genes of their cliff-dwelling and ground-feeding ancestors, and by selective breeding, they do not even retreat to sewers, trees, or parks to defecate, mate, and live, as do so many other animals. Further, these birds may evoke discomfort or even nausea by scavenging humanity's refuse. Their metaphorical "pollution" of city streets becomes crystallized through their link to humanity's literal pollution—trash. Part of our aversion to pigeons, then, stems from cultural insecurities about proximity to dirt and impurity. Mary Douglas argued, "In chasing dirt, in papering, decorating, tidying, we are not governed by an anxiety to escape disease, but are positively re-ordering our environment, making it conform to an idea."[56] The appearance of "dirt" where it does not "belong," whether a soiled white shirt collar or pigeons on a sidewalk, is taboo, eliciting a reflexive human desire to remove the pollution and thereby restore a sense of moral integrity.

Pigeons have lived and scavenged in cities for centuries. But over time an increasing number of urban spaces have been redefined as off-limits to them, often marked by plastic spikes and by the words "Do Not Feed the Pigeons." *Columba livia* is now a "homeless" species, surviving in the urban interstices off of society's occasional generosity and its refuse. As if the link to their human analogues were not clear enough, public discourse and the media are filled with references to pigeons as "bums" and "squatters." And, as New York and other cities made efforts to "clean up" their streets to enhance the "quality of life," both homeless people and pigeons came under increased scrutiny and symbolized the type of dirt and disorder that officials sought to rein in.[57]

Just as the pigeons were tied to broader problems in London and Venice, their solutions entailed broader agendas. To enforce the ban on pigeon feeding, both London and Venice hired security forces to patrol the squares. These "stewards" (Venice) and "heritage wardens" (London) do not only enforce the feeding ban, though. They police a wide range of activities that either have been recently criminalized or were on the books but rarely enforced. In Piazza San Marco, they now ensure that visitors do not consume food on the square or sit on the steps. While this law was on the books when I was there, it was rarely enforced. The result of enforcement is that, combined with the ban on pigeon feeding, if one wants to linger in the square the only legal options are to stand around or sit at the exorbitantly priced cafés (where an espresso will set one back about $14). This is no small matter in a city that has few public spaces aside from the piazza. In Trafalgar Square, it is now impossible to escape the gaze of a heritage warden for long. One nostalgist described a visit to the new Trafalgar Square as like "walking across a prison exercise yard only to spend a week's wages on a cup of foul coffee" and be harassed by the "pigeon police," and a newspaper columnist labeled the pigeon eviction and heightened surveillance "dour Cromwellian austerity."[58] One common form of interaction that I witnessed was heritage wardens telling visitors that they were doing something wrong. Some of these used to be activities that "the first secretary of state considers not to merit criminal sanctions and on which he will not therefore normally be prepared to confirm bylaws.[59]

While Piazza San Marco and Trafalgar Square are significant historical places worthy of protection, the measures implemented to control behavior in these spaces seem to constrain the possibilities for how visitors can form a personal attachment to them. Though the monuments may benefit, banishing the pigeons and heightening surveillance may exorcise the ghosts of these places whose presence was so enchanting to many visitors. The mayors of London and Venice portrayed pigeons as a disruption of the social order, but many folks viewed the birds of Piazza San Marco and Trafalgar Square as inimitable quirks of place—unlike any other pigeons—and experienced feeding them as a way to consume local culture. Thus, while a large concert with corporate sponsorship brings large crowds to Trafalgar Square at punctuated intervals, it may do so at the expense of the spontaneous and idiosyncratic interactional moments that made this space a place *like no other*.[60]

The pigeons of Trafalgar Square and Piazza San Marco are objectively different than most other street pigeons—they have been tamed. In the process, they have become fully dependent on people for food and have stopped scavenging. Cultural products in a literal sense, their habits and their large flock sizes are a result of ritual human practices that engender a tradition. Most of the birds cannot survive without interacting with people, leading animal rights activists in both cities to ask the thorny question of whether people owe anything to these animals that they have made into urban scavengers. Once granted a space and a food supply, do these animals deserve a place in the city? Through the pigeons of Piazza San Marco and Trafalgar Square, we can understand how people interpret, construct, and contest the meanings of these and other urban milieus. This process also reveals how nonhumans can enchant the built environment in ways that are different than those often highlighted by environmental scholars. The ubiquitous pigeon, so often the scourge of city streets, enchanted these squares for so long not primarily because it symbolized nature but rather because it made possible a felt connection to culture.

II

The Totemic Pigeon

THREE

New York's Rooftop Pigeon Flyers

Crafting Nature and Anchoring the Self

WHILE I BEGAN MY research looking at the place of feral pigeons in public space, I became curious about the huge stocks of domesticated pigeons I saw swarming in tight bundles over the streets of Bushwick, Brooklyn, where I lived. From my roof, I could regularly discern six or seven men on top of tenements trying to choreograph their birds' movements by whistling and waving long poles. To get to know these "pigeon flyers," in April 2005 I started making regular visits to a nearby pet shop named Broadway Pigeons and Pet Supplies. Joey Scott, the affable Jewish Italian owner, agreed to introduce me to his regular customers. Over the next three years, I spent almost every Sunday at Joey's store, getting acquainted with 44 regulars and visiting the coops of half of them. The flyer I got to know best was Carmine Gangone, an old-timer who had been flying pigeons for 75 years and whose coop was located by the 80th Street stop of the A train in Queens.

The five boroughs' once ubiquitous rooftop coops are now as scarce as the Italian longshoremen known for building them, portrayed in iconic representations of working-class New York like the 1954 film *On the Waterfront*—where Marlon Brando was either working on the docks or flying pigeons on his roof. Where once, Carmine lamented, his block hosted five or six coops, he now needed binoculars to spot other flyers. And while Carmine viscerally recalled the days when the area, known as Ozone Park, was an Italian enclave where generations of a family lived on the same block, Puerto Ricans and immigrants from places like India and Guyana had taken their place.[1] Though Carmine was apt to eulogize his neighborhood's blue-collar Italian character, he chose to age in place rather than follow his family to the suburbs.

Because pigeon flying was historically the domain of working-class white men who passed on the practice, and their coops, to their sons, the number of flyers declined precipitously over the second half of the 20th century as many upwardly mobile whites migrated from New York's outer boroughs to the suburbs. But pigeon flying is not dead yet, and by making the four-mile trip to Joey's pet shop, Carmine got to socialize with other elderly and middle-aged Italians who commuted in from more genteel neighborhoods like Carroll Gardens and Bensonhurst. Carmine also mixed with young and middle-aged Hispanic and black men who flew pigeons in the immediate vicinity of the pet shop. These men reflected a newer cohort of flyers that picked up the hobby from ethnic whites as kids when they moved into neighborhoods in transition such as Bushwick and East New York. The most famous New York flyer of them all is in fact black: Mike Tyson. On my first visit to the pet shop, Joey was quick to flash a picture of himself with the former boxing champion and brag to me in his thick Brooklyn accent, 'We're friends!'[2]

The racial composition of the men who frequented Joey's pet shop was about two-fifths white, two-fifths Hispanic, and one-fifth black; flyers from more varied backgrounds did pass through, including Middle Eastern men. The whites were overwhelmingly of Italian origin (though American born) and were retired (their mean age was about 70). Most of the Hispanics were Puerto Rican, and most of the blacks were African American. Nearly every man of color was middle-aged or younger and labored for a wage (their mean age was about 45).

While chapter 5 documents pet shop interactions, this chapter depicts the rooftop experience. The flyers intensely enjoyed breeding, raising, and handling their birds and watching them in flight. Though some environmental scholars view such intimate relations as resulting from an innate human drive to associate with other species, or as enabling the transcendence of social life, I saw how flyers' appreciation for their pigeons was given impetus by their social relations: they crafted "purebreds" and evaluated their performance based on established group customs, and they measured their birds' worth and their own status as a flyer against each other. Additionally, their coops were nodes through which they experienced and connected to their neighborhood. For the younger men of color living in areas still heavily populated with coops, they interacted with each other daily through their birds, coordinating times to fly their stocks and catching each other's pigeons. For the older white men, their coops were like time capsules, linking them to their long-gone romanticized ethnic neighborhoods and mitigating the impact of the changes around them. Flyers also spent hours each day doing gritty work like "scraping shit" off their coop floors. Instead of complaining about this labor, I found that most men accepted and took pride in it because it was a way to affirm and perform their working-class sensibilities.

Rather than cultivating a sense of connection to nature, the flyers' relations with pigeons embedded them in a distinct social world. And because each flyer sculpted a unique bloodline of pigeons and was granted or denied status by his peers based on his birds' appearance and performance, his stock became an extension of himself. This shows one way that the *social self* can be constituted through interactions with animals. The men also illustrate how people's relations with nonhumans are shaped by social categories like class and gender. These observations are a caveat to sociological perspectives that assume nonhumans do not play a consequential role in organizing the social realm, and to sociobiological or ecological theories that emphasize the presocial determinants and asocial rewards of humans' relationships with animals.

Carmine and the Old Guard

I met Carmine, a sturdy Italian American World War II vet in his mid-80s with tousled gray hair and blue eyes, on my second visit to Joey's store, when he came in with his "partners" Frankie and Charlie. "Partner" was the label flyers used to formalize the working role of someone who shared rooftop responsibilities, as Frankie did with Carmine. Men sometimes also used the term simply to signal a close friendship with another flyer, as Carmine did with Charlie. The men's jointly shared bands that they placed around their pigeons' legs, which read "C.F.C. 80TH STREET OZONE PARK," inscribed their relationship. As Joey placed four sacks of feed on a dolly at the direction of Frankie, a stout balding Italian American man of Carmine's vintage with sunken brown eyes who chain-smoked in the store, Carmine and Charlie headed to the back of the shop to inspect the pigeons for sale. As Carmine entered one of the pens and the panicked birds flitted around him, he focused his attention on a brown speckled pigeon. He then quickly extended his right hand and snatched it from above, pinning its wings with his strong stubby fingers. Carmine held the bird to the chicken wire while splaying its wing and said, 'That's a nice pigeon, isn't it, Charlie?' Then he turned the head of the now docile pigeon so he could examine its eyes and nose. He ran his fingers down the tail and signaled to Charlie to put it in the holding pen. It was a good bird. As he scrutinized the other pigeons one by one, Carmine complained about how they had "bulleyes" or "splashed beaks," imperfections that fell short of his strict standards. Charlie seemed to be trying to absorb it all, asking, 'What about that one?'

The pigeon was going home with Charlie, a gentle and stocky Puerto Rican man in his early 50s with short curly hair and glasses. He lived, and flew pigeons, near Carmine and routinely gave him a ride to the store. Carmine appeared to be his mentor, whether Charlie liked it or not. When Charlie showed interest in a brown and white pigeon, Carmine groaned, 'That bird is no good!' Charlie protested, but Carmine shot back, 'Don't tell *me* about birds! For Christ's sake, look at the wings! See this feather? It's *supposed* to be all white! This bird is a cross [breed]!' Smiling wryly, Carmine added, 'Take it, it fits right in with the garbage you got!'

After Joey introduced us, Carmine immediately began teasing me. I had long, unkempt hair (dreadlocks) at the time, and Carmine grabbed it as he berated me with his unmistakable Italian-infused Brooklyn accent, 'Jesus, how did you let this happen?'[3] Joey (thankfully) interceded, 'He's writing about the birds. I told him you're one of the oldest flyers around here.' Joey turned to me, 'Carmine's the *best*.' Carmine invited me to his roof right away. For the next three years, I would go to his house every few weeks and spend half the day on his roof and in his kitchen. Carmine treated me in an affectionately paternalistic way, sometimes asking if I needed money and chiding, 'When are you gonna start eating meat? You're too skinny!' Frankie, a bachelor who moved in with Carmine in 2003, ensured that I was well fed.

Carmine shared the first floor of his two-story townhouse with Frankie and rented out the top floor. A dark hallway with old fake-wood paneling—hardly lit by a bare bulb—led to his modest home, which to me felt untouched since the 1950s. His kitchen walls featured pictures of his sons, grandsons, and great-grandsons (most of whom lived in Long Island), as well as a hand-colored photograph of him and his now deceased wife on their wedding day.

On my first visit with Carmine, a sunny spring day in May 2005, we immediately went to his roof. After climbing the stairs to the landing, Carmine led the way up a metal-rung ladder and though a hatch. The screeching elevated A train, just a half block away, assaulted our eardrums. From the flat tar roof, I could see for miles in every direction. Planes descended into JFK airport, steeples peeked out above treetops, and satellite dishes adorned the uniform homes. After the train left, I could make out the squeaky cries and bass-heavy coos of baby birds and their parents emanating from two simple coops, made of plywood and two-by-fours, spaced about 15 feet apart. Small windows and screen doors made the coops look like mini houses. Attached to the right side of one coop was a small rectangular box known as the "prisoner's pen," in case Carmine caught any "stray" pigeons from other flyers. The dozen or so colored leg bands curled around the screen were little trophies of an occasional score.

As Carmine filled up the water canisters, he recounted his introduction to pigeons. "I had birds since I'm five years old. I had *one* pigeon my older brother brought me. He found it in the street. I didn't know *nothin'* about birds." Carmine laughed, "I liked it. I put it in a cardboard box—in the house. My mother went *crazy*. Then the next morning my brother's down in the yard, building a coop for me. When he got paid, every Friday he'd go to Maspeth [Queens] and buy birds. And before you know it I got a stock down there. This was in Brooklyn. I've had them ever since. The only time I didn't have them was three and half years I was in the service."

Both coops were lined with wooden one-foot-square cubbyholes. When I entered the cramped "breeder's coop," dozens of pigeons jumped out of their tiny homes and circled me frantically. These prized birds, which existed solely to beget offspring, were never allowed out. About half of the three dozen or so boxes had cardboard bowls in them, some containing bald, awkward-looking babies with prickly quills that would become feathers. Carmine explained that the larger coop was where his "flying stock" was kept. "These birds are out every day, winter or summer." He flew his "young ones," pigeons bred that year, separately from the older birds so that he could train them to fly as a stock and compare their performance to previous generations.

As we peered into the coops and dozens of beady eyes peered back, Carmine explained the breed of pigeon that he kept. "These are all *flights*. A flight is a solid [colored] bird with white tips, see?" Carmine's birds varied widely in color: white, black, chocolate, dun (light brown, silver accented), strawberry (dilute red), and yellow. They were slightly smaller and nimbler than feral pigeons, their eyes were pure white, and their beaks were flesh colored and thin. Traditionally, flights should be a single color except for white wing tips. The ideal standards for the flight, which is an official exhibition breed, have changed remarkably little since they were

codified a century ago. But, among the New York flyers, a checkerboard-patterned variety called a "teager" and a barred-wing variety known as an "Isabella" had become accepted subtypes. Other agreeable local modifications included "caps," in which the feathers on the back of the head turn upward, and "beards," in which a splash of white hides under the beak. The subtle genetic markers that produce variations in color and pattern were the building blocks with which flyers constructed their own signature stock of pigeons.

By keeping flights, Carmine situated himself within both an Italian and a New York tradition. "The American Domestic Flight," pigeon expert Wendell Levi wrote, "is an American creation, which has been . . . almost exclusively bred in the City of New York" since the mid- to late 1800s. It is also known as the New York flight. Levi expounded, "Breeders vied with each other in capturing other fellows' birds. The underlying principle was the same as that of the *triganieri* in Modena [Italy]."[4] This thieving competition dates back to the 1300s in Modena, but when Italian immigrants brought the game to New York, they created a new foot soldier by crossing several German breeds.

While most flyers I met kept flights, a minority favored tiplets, an English breed created for endurance contests. Carmine said he preferred flights to tiplets because "they do more for you." That is, they can be trained to "route," or fly long distances and return. "See, a tiplet won't do this. They don't have that instinct to move out. They want to fly just over the coop. But these things [flights], I let 'em out and all of a sudden they take off. Either they go to Jamaica [Queens] or go way into Brooklyn. And you don't see them! That's my pleasure. And to see them come back, without losing one—it's amazing! A lot of guys are afraid to do it! [They] tried routing and they lost too many birds."

Carmine's enthusiasm for flights was also a moral matter. For instance, he regularly berated Charlie for having "garbage" birds. Yet in this context, "garbage" did not mean birds that failed to live up to institutionalized standards of a breed. "Garbage" meant any birds that were not flights. One afternoon in Carmine's kitchen, Charlie explained to me that he did not like to keep the same birds around, and that he enjoyed the challenge of training various types of pigeons—such as tiplets and German owls—to fly as one stock. Charlie added, 'It should be a matter of taste, the kinds of birds you keep . . . ,' but Carmine cut him off and exclaimed, 'Those birds should *not* be in the city! This is New York for Christ's sake, we keep flights!'

Carmine adhered to a strict schedule with his birds, from which he seemed to derive particular pride. On a chilly September morning, I arrived at his house while it was still pitch black. The stale coffee he served me at 5:50 indicated that he had already been up for some time. On the roof, Carmine opened the flyers' coop and chased the birds out by walking inside. Simply by whistling, he made the 150 or so "young ones" simultaneously take off into the dark blue morning sky. They went high, and Carmine remarked how much energy they had. They made sharp turns in unison, looking like a tornado. Occasionally they would circle low and fly between us on the roof, their flapping wings emitting a squeaking sound. Carmine shouted, 'Come on, climb you bastards!' Then he whistled, picked up his bamboo rod with a trash bag attached to the end, and began waving it. In teaching his birds to stick together, Carmine built on pigeons' natural inclination to fly as a group, which makes it harder for predators to hone in on one bird. The bamboo rod startles pigeons, triggering this collective response. At first, the birds rose but split into two groups. When they merged again after he whistled, Carmine shouted 'Yes!' The sun finally emerged, catching the pigeons' white tips as they swirled.

After "chasing" the birds until they finally moved out in the direction of downtown Brooklyn, Carmine got to work. He dumped the watering

Photo by Marcin Szczepanski.

cans down the gutter, hosed off the feces, and refilled the cans. He checked on the pigeons in the breeders' coop. Then he took a metal hoe and went into the flyers' coop to scrape off the feces while the birds were still in the air. 'I want you to be honest with me,' Carmine pleaded. 'Have you been able to see the floor of the coops you've visited? My God! Some guys let their coops get so filthy!' By the time Carmine hosed hundreds of dried fecal stains off the roof, it was 9:00 and had turned into a warm sunny day. Carmine remarked, 'These are the kind of days that would make me want to stay on the roof all day, boy. I would miss appointments sometimes! And my wife would bring me lunch.' He complained, as he often did, about the lack of "action" nowadays. 'I wish I had stocks around here to tangle with, but they're all gone.' The lack of action stimulated Carmine to train his birds to route. This way, he could still 'have some fun' even though he did not get to "crash" other stocks. Repeating a mantra of flyers residing in "dead zones," he remarked, 'You gotta create your own action.' A hawk attack could also bring action: 'I love it. When the hawk comes is when you can really see what your birds are made of. To watch 'em split in every direction and come back, without losing any, is something else.' Carmine tossed scoops of grain on the roof, and Frankie emerged from the hatch with a cigarette in his mouth as we scanned the leg bands of his returned birds to see if there were any strays among them.

There almost never were any strays, but Carmine's and Frankie's constant search for them was grounded in habit and redeemed just enough by the occasional find to justify checking. The practice also offered opportunities for reminiscing about the good old days, experientially connecting their current practices to a distant time and place in which there were not only more pigeon flyers around, but also more family and friends. Carmine recalled, 'You used to be able to catch like 20 or more strays a week and sell them to the pet shop. You wouldn't even have to pay for feed!' Frankie added, 'There used to be like six pigeon guys on this block. I used to catch strays without even looking.' They both scanned the sky, and Frankie said he saw pigeons miles away in Brooklyn. 'Where?' Carmine asked. 'It's a mirage!' But binoculars confirmed that it was a stock. They debated whether it belonged to "Tony the Terminator" and fantasized about a hawk attack that would scatter the other stock and cause strays to land on Carmine's roof. This led to an extended discussion in which they inventoried perhaps two dozen long-gone flyers that used to live nearby. Carmine and Frankie commented on almost every bird in the sky as they sized up whether it might be a stray, a hawk, a gull, or perhaps just a "street rat."

This was how Carmine spent his days. After he finished his morning routine on the roof, he sometimes visited a nearby park where he had befriended several young black and Dominican women who cleaned it. But if the weather was nice, Carmine could not stay off the roof for long. At noon, he would change the water cans again and feed the pigeons. If it was breeding season, there was a lot more work to be done. The birds had to be given grit to make their eggs' shells harder. They ate more, so they defecated more, and so Carmine had to clean a lot more. Also, Carmine needed to place an identifying band over all of the babies' legs several days after birth. Frankie usually came home from playing cards at an Italian social club around 2:00 and made lunch for them (Carmine admitted, 'I can't cook a damn thing!'). He would then join Carmine on the roof, where they continued their ongoing discussions about various birds. Frankie might ask, 'Where's that grizzle cock we caught the other day?' Or Carmine might say, 'Is that dun teager hen still sick?' Routinely, they singled out a bird from hundreds and recounted its history: 'We bred that two years ago out of those two silver caps.' They also closely monitored the stock's health, quarantining and medicating sick birds. After breaking again for coffee, Carmine would return to the roof and persuade the pigeons to fly once more. As the sun slipped away, he would rest on

a small bench and maybe nod in and out of brief sleep as the birds also wound down, landing and lazily preening themselves. Finally, he would look at the leg bands to identify any strays, toss feed into the coops, and lock them up.

'Every year I say I won't raise young ones and route them,' Carmine once laughed as he massaged an aching shoulder. 'But what would I do if I didn't do this?' As Carmine was wont to do, he simultaneously complained of the absence of pigeon flyers and changes in the neighborhood. 'Everybody around here is Indian or Muslim. Don't get me wrong, most of 'em are OK. But they're not my kind of people.' He added, 'Oh Christ! There were birds all over the place, and now I'm the only one left.' Sighing, Carmine concluded, 'But I can't leave now. Maybe if I was younger I'd get out. But now what's the point? You go somewhere to die?'

Carmine interpreted the city and adapted to its changes from his roof. The minds of old flyers like him were layered with mental maps of a vanished city organized around their pigeons. They pictured their neighborhood as it was decades ago. They enumerated the names and locations of long-gone flyers. They recalled a clear view of the skyline now obstructed by new buildings. And they conjured up addresses of the many pet shops that once dotted the cityscape and told endless stories about the colorful personalities that haunted them. As new, anonymous condos went up or strangers speaking foreign tongues moved in, Carmine and the remaining flyers of his era found that their rooftop experiences were transformed. Where once thousands of pigeon wings filled the air—directed by members of the same ethnic and age cohort as Carmine—now silent skies reigned. But the pigeons also provided a balm that soothed the malaise resulting from such drastic changes. On the roof, life went on and the passage of time seemed to slow down. Hawks were the eternal antagonists. Lasting, Platonic ideals guided breeding. When Carmine was immersed in caring for the birds, his body and mind fell into ritual habits that still produced satisfying results—and satisfying relationships. Few innovations in training or medicine had come along to challenge his experiential authority. Such repetitive mundane practices formed a tangible, unbroken chain between past and present.

Though Joey Scott was less than half Carmine's age, his biography traced Carmine's and other old-timers'. Joey, 37 when I first met him, was average height and weight with a bald head and blue eyes. When working at the pet shop, he usually wore a white T-shirt, sneakers, and faded jeans. Raised in Canarsie, Brooklyn, when it was a blue-collar Italian enclave with a sizable Jewish population and a number of pigeon coops, Joey followed in his Italian grandfather's footsteps (his Jewish father never kept pigeons). He was a truck driver before opening the pet shop with his older brother Michael in 2002, and remained the store's sole employee, working six days a week, until 2007. Then, with a newborn son and the store barely breaking even, Joey took a job in construction but continued to work in the shop on Sundays.

Joey lived in his deceased grandfather's detached home with his grandmother and his Puerto Rican girlfriend and their two children. As his neighborhood changed in ways that might have made it unrecognizable to his grandfather, he said that he became more withdrawn. "My neighborhood is no longer Italian anymore," Joey explained. "There may be a handful of people left there that I even know. And I lived there 38 years. Everybody I know moved or passed away." Census figures underline his experience: between 1990 and 2000, the white population of Canarsie fell from 70% to 35% while the black population increased from 20% to 51%.[5]

In Joey's backyard sat his grandfather's original plywood coop. Joey still used it for breeding, but the alarm system and padlocks were his own modifications. Though the architecture and greenery of Joey's block evoked a villagelike gentility, a few blocks away the commercial street near the L train subway entrance—with its decrepit storefronts, liquor stores, and check-cashing kiosks—was an indicator of the declining economic fortunes of the area. Joey built his own coop on the roof, complete with a skylight and sky-blue vinyl siding, where he kept his flying stock. After work, and after enduring the hectic five-mile rush-hour commute home, Joey would escape to his rooftop. As the birds flew around the church steeple across the street and he performed routine chores such as spraying down his immaculate coop with a hose, the scene had all the trappings of suburban domesticity.

There were few opportunities anymore for Joey to tangle with other flyers. While Joey said, "I don't fly birds to catch birds," this seemed to reflect the reality that there were few flyers around rather than a preference. "Where I am, there's really not a lot of catching strays," Joey explained. "Over there in Bushwick, every other block has a lot of birds. They're the guys looking to catch each other's birds . . . I just fly them for my pleasure." Joey admitted, "Maybe if I had somebody near me who flew the birds I'd feel differently." In fact, on the rare occasions that another flyer's stock did make a pass by Joey's roof, he vociferously attempted to catch strays, and he enthusiastically recalled the glory days when the "pigeon wars" raged around him.

As we watched Joey's stock of flights and tiplets circle in front of the setting sun one afternoon, I asked him what his favorite thing was about flying pigeons. 'I just like being up here, period,' he replied. After a moment of silence, he added, 'And I bred them all. Every bird up here I bred.' Indeed, many flyers agreed that Joey had the finest short-faced tiplets around. Joey's days on the roof, like those of Carmine and others who saw no "action" in the sky, were times of relative solitude. They adapted their activities accordingly, taking a little more care than they used to in mating up birds for breeding, and finding pleasure in the simple things. As we watched Joey's birds drop in for a landing one by one, he smiled and remarked, 'That's phat the way they come in like that and pull up to stop.' Joey imitated the maneuver, using his folded arms as wings. He then locked the birds up and went downstairs to join his family for dinner.

Panama and the New Breed

In the neighborhoods encircling Joey's pet shop, things were different. A one-mile stretch of Wilson Avenue in Bushwick, for example, featured over a dozen rooftop coops. The men who commanded the birds in these areas, like Panama, tended to be middle-aged or younger and nonwhite. Panama, whose real name is Delroy Sampson, got his nickname from his country of origin. A muscular black man in his 50s with cropped hair, he migrated to a predominantly Italian section of Bushwick when he was a young child. He said he spoke no English when he arrived, but he no longer had a discernible accent and passed for African American. Panama was smiling, outgoing, and confident. He winked as he boasted about his birds' performance, or the performance of his teenage daughter on her standardized tests. He regularly wore fingerless gloves, a thick gold chain with a medallion, and a bicycle hat (one said "Brooklyn" on the brim while another was emblazoned with "USA").

A renter when I met him, in 2006 Panama bought a newly erected townhouse in an "up and coming," but still economically depressed, section of Brooklyn known as Broadway Junction (next to Bushwick).[6] A ceramic decoration of two pigeons hung on the front door. Panama was an electrician, and said he had been one for 35 years. In one of my visits to his house, Panama seemed to brim with pride as he showed off his certificate that declared he was a foreman in the local union and detailed how he had consolidated his loans, borrowed against the equity of his home to renovate the basement, and saved up to buy a new Chevrolet SUV.

Sometimes Panama hinted that his hold on his lifestyle and respectability was precarious. He told me that he made sure to get two copies of his foreman certificate, which he stored in different places, so that he could never be caught without it if someone challenged him. On my first visit to his roof, he said, unprompted, "I'm trying to do a family thing. Not get caught out of position in some bullshit. I'm doing my business, everything is looking good." Panama accepted the burdens of being black in America and tried to model his behavior accordingly. As images of the star NFL quarterback Michael Vick in handcuffs flashed across the television screen after his arrest for dogfighting, Panama complained, 'He didn't kill nobody. But he's black.' He shrugged his shoulders, 'We have to hold ourselves up to a higher standard.' This launched him into a story of how he went to Puerto Rico in 1997–98 for a job but 'was getting paid crap' because he was black. He recounted that he returned to New York, found a well-paying job, and sent the pay stubs to his former job site to prove his worth. Panama brought this sense of dignity and determination to everything he did, including his pigeons.

The story of how Panama got into pigeons is an unusual one. He recounted, "When I was ten I saw a Walt Disney presentation, 'The Pigeon That Worked a Miracle.'" I started to laugh, and Panama chuckled but seemed defensive. "It was some serious shit!" In the movie, a boy becomes paralyzed but the doctors suspect his injury is psychosomatic. He takes up pigeon racing, and when a cat tries to pounce on his winning pigeon, the boy leaps from his wheelchair, saving the bird and curing himself. Panama said that the movie inspired him to begin hanging around a homing pigeon club in nearby Maspeth, Queens. There he befriended the president Joe Larocco, who convinced Panama's parents that pigeon racing would help keep him off the streets. "Ever since," Panama grinned, "I've been flying pigeons." He started out racing homers but eventually transitioned to flights, like those around him in Bushwick.

Panama built his rooftop coop, roughly 12 by 6 by 6 feet and painted red, out of drywall to meet fire safety standards; and he said he paid $3,000 for a building permit. In the year before its construction in 2006–7, Panama routinely showed off his blueprints in the pet shop. After we ascended the ladder to the roof on a warm September morning, he unraveled his crisp, large Panama flag and placed it on the coop. After putting smooth jazz on the radio, which helped drown out the sirens and screeching brakes of the nearby elevated train, Panama opened up the screens so that his 350 or so pigeons could fly. As they stumbled out, he turned to me and said, "My birds fly nice. They say Panama talks a lot of shit at the pet shop, but I back it up."

Panama whistled, and all the birds took off at the same time, their wings thunderously smacking together in unison. He bragged, "See that? I don't even gotta move. I got 'em on fingertip control." The birds quickly circled upward and outward. "They're high already," I remarked. Panama responded, "Yeah, but that ain't nothing." He grabbed the flag and waved it fiercely to chase the birds higher. "In the evenings they cut the fuck *out*. They were gone at 3:00 yesterday and those bitches didn't come back until like 6:00. They were *gone*, man!" Panama admitted he lost some birds by "rolling out": "Manny caught two; Joey . . . got two; Terry dropped one but didn't catch it; and his boy over on 88th Street caught seven. So that's eleven, and that's it. I tell 'em, OK, I lost a few. You know, they screaming over shit. Don't mean nothing to me." Anyone who rolls out in this area, Panama believed, must expect to lose birds.

Once the birds were up, Panama changed the water cans, exclaiming, "As many times as they dirty the water is as many times as I change it. They going into a crystal pool." Then he began scraping the feces off the floors with a hoe and sucking up feathers with a wet vac, which he did every other day. "I keep my shit clean!" Smirking, Panama complained, "When you go by some of these guys, they don't do *shit*! Their coop is dirty, birds are dirty." Pointing toward his medicine cabinet stocked with vaccinations and herbal remedies, Panama added, "You can't expect to have good flyers if you can't take care of what you got."

Panama had to chase the birds every day to get the performance he wanted. The immediate goal was to get them to fly high and stay together and, just as importantly, to avoid going down on other flyers' rooftops. "If you want 'em to fly anything decent, you gotta punish 'em in the morning. You gotta make 'em understand they're not going back for breakfast or going back to bed." Yet Panama conceded that the birds' performance could be unpredictable: "It has to be the right day to get what you want." Weather and season were two key variables, as pigeons molted in the fall, overheated in the summer, and disliked flying in the rain. Today was not an optimal day, as it was hot and sticky. Panama complained, "They not gonna push. Those flyers give me a hard-on when they just *go* for hours. I wish the hawk would show up and give 'em a fuckin' beating. I want them to go left to right, and then go. But they gonna just do a downward spiral. You can feel the day."

Contradicting his braggadocio about having "fingertip control," the pigeons reminded Panama that they were living creatures with their own wills and limits. Though he pushed the pigeons, they shaped his expectations and flying schedule. Their refusal to roll out signaled to Panama that

Panama displays his medicine cabinet.

he had to wait for another day or allow his pigeons to rest. He and his birds developed a routine that, in the long run, accommodated both parties. The pigeons pushed off as soon as Panama unfurled the flag because experience taught them that he chased them with it. However, if they repeatedly returned to the roof, flew low, and landed with their mouths open (which signaled that they were hot and thirsty), Panama knew from their shared history that the birds had reached their threshold and that pushing them was useless. Indeed, extending the birds' natural limits and reining in their "individuality" was a fundamental goal of flyers.

While Panama recognized the pigeons would not always perform, he had his limits: "The birds that come down early I sell to the pet shop. I don't care how pretty they are. Then what you have is a powerhouse. And when you roll out and mix somebody, my birds go *in* 'em, and *out*. They leave." To emphasize his point, he said, "Look over there, that white face [pigeon]. He's pulling out. He don't pull out no more. He's not here." Panama planned to get rid of him. "When they go against the grain . . . you're not getting what you want."

It was important to Panama that his disciplined pigeon army was the product of his own hands: "You have to know what you have. These guys are bringing in birds from everywhere. I'm not adding nothing to this. I'm gonna cut *out* birds, make it tighter, and breed what I already got here that shows me intellect and knowledge . . . in the course of the year." In this way, Panama created a "power team" uncorrupted by poor performers or foreign birds. This bloodline became Panama's signature, marking his birds as unique from other flights and entangling Panama's reputation in their DNA. When other flyers caught Panama's birds, it was as if a minuscule part of Panama had been captured.

On a daily basis, Panama interacted with flyers around him through his birds. He pointed out three coops within five blocks of his house: "Sean flies about 600 there. And there go Louicito . . . " Just then, his Nextel walkie-talkie crackled. It was Sean: "Yo, that your birds rippin' it up there?" Panama laughed, "Yeah, that's mine." Sean yelled back, "I'm waiting for 'em." After he hung up, Sean pushed his stock up, and we could see his bundle approaching Panama's stock. The phone chirped again: "Yo! It's me, Big Dog in the house! Send 'em over to me so I can catch a nice [bird]." As we watched other stocks, Panama commented on the flyers just as they were surely sizing him up: "Louicito has been flyin' good." Panama called him: "Hey, Louie. I'm watchin' 'em fly. Oh, you watching *mine* fly? I got the professor up here and . . . I told him your birds are flyin' pretty good. Did you catch anything?" Louicito responded, "I didn't catch nothing."

This is how Panama passed the hours before and after work, as the birds' interactions in the sky spurred connections with flyers on the ground. Whereas Carmine and Joey lived in "dead zones," Panama had at least five coops within a ten-block radius that he could tangle with. These aerial encounters motivated Panama to always "come correct" with his birds. The worth of his pigeons—and by extension, his status among the flyers—had to be repeatedly demonstrated in the sky. In this way, Panama's pigeons structured his social relationships. He and his neighbors developed routines in which they expected each other to "put up" their birds and risk losing them. If Panama passed a day without putting up his stock, his peers got on his case. The men regularly caught strays off each other and exchanged birds: "The guy who called me before, Carlos, had caught a nice yellow magpie. And he wanted a nice bird for it. I didn't have it at the time. I told him, give it to me and I'm a give it to the kid [Louicito, who flew magpies] and I'll catch you a nice bird. The next morning, I caught a beautiful Isabella flight with a beard and a cap from Terry—the sanitation guy—and I took it right to him."

Panama said he mentored Louicito and returned captured strays because he was only 19 and new to flying. "He's learning from me now. If my birds mix him, I'll give 'em back. I'm gonna keep souping him up so that he makes 'em fly." But when it came to other flyers, it was a different story: "I got Sean who flies up there—he got about 600—and we *killing* him. He's scared to put up now! Even my daughter's got him terrorized to put up!" He laughed, "He's threatening me, I'm a do this, I'm gonna do that—and I got 300 birds to his 600 and I'm catching 5, 9, 6. I'm up to 37." I asked Panama if he liked the competition, and he grinned like a Cheshire cat. "Yeah, of course. *I live for the day.* I say, *I live to roll.*" Like many flyers, Panama did not give back most of the pigeons he caught. Yet, unless they were particularly fine specimens, he sold the strays to the pet shop for a few dollars. What the men ultimately did with the prisoners was less important than the act of catching them, like sport fishing.

Like Carmine, Panama experienced his neighborhood through his roof. He too had a mental map of the city based around his animal practices—but it was a contemporary map. His pigeons grounded him in the here and now, not the past. Panama was surrounded by people who looked and spoke like him, including a half-dozen flyers within striking distance of his coop. He could cycle to Joey's pet shop. The majority of the flyers of color I came to know fit broadly into Panama's social situation. Guided toward certain segments of the city both by external constraints and by ethnic attachments, they recreated a robust concentration of flyers—though smaller than the heyday of the Italians—spanning the environs around Bushwick.

Another man whose biography and rooftop experiences echoed Panama's was Richie Garcia, a youthful and jovial 50-year-old Puerto Rican man with cropped hair, a neatly trimmed goatee, and wire-frame glasses who often sported a New York Yankees baseball cap. He immigrated to Bushwick with his family when he was 13, and now he flew his flights on the roof of a five-story tenement that he owned and rented out on Humboldt Street in nearby East Williamsburg. As a teenager, Richie said he used to sit in Bushwick Park and watch the old Italians, who dominated the area at that time, fly pigeons on their roof. Over time, he convinced them to let him up onto their rooftops, "and the next thing I know I am shoveling shit for the rest of my life!"

At 3:00 on a sunny October afternoon, I arrived on his rooftop to find Richie waiting to see if a flyer with a coop about ten blocks away would chase his stock. "Maybe we'll have some fun today." At 4:30, the man put up his stock of about 200 tiplets. They circled like a swarm of bees, and each rotation brought them higher and stretched out their orbit. Richie stared up in anticipation: "Come this way and see what happens!" Then another flyer near the elevated train put up a small stock of homers, a rarity in this area. It seemed a clash was imminent between the two stocks. As the homers approached, we watched the man chase his tiplets higher into the sky with a bamboo pole, hoping to bring the homers to him.

The stocks of pigeons approached each other as a unit, but they retreated just before making contact. Yet the flyers were whistling and discouraging the birds from turning back home by chasing them. The tiplets finally circled around and collided with the stock of homers. Richie narrated: "Boom! They're in there. Oh, look at that shit! Right into 'em!" A push and pull ensued in which the birds broke up into two bundles but then merged again, like magnets, likely attracted by the same "herd instinct" that flyers relied on to train them. It seemed they kept trying to separate into the two original stocks, but as they did, some birds realized they were in the wrong group and swapped sides. As this cat-and-mouse game continued, a gull appeared and "spooked" the pigeons. The two bundles dissolved as pigeons scattered in every direction. Richie ran to grab his bamboo rod. This was his chance to catch strays. "Oh! They comin' this way, bro!" As a small bundle flew above our heads, Richie clapped, tossed his Yankees cap, and ferociously waved his bamboo pole to coax his 300 flights up to the strays about 200 feet over us: "Come on! Get up there!"

Richie's pigeons "mixed" the bundle, but it flew past and circled back. The stock of tiplets regrouped and formed a moving wall that extended from the roof into the clouds. We could see scattered birds finding their

way back home from several directions. As the golden sun faded, Richie brought his stock down by tossing seed on the roof and then scanned their leg bands, revealing two captured strays. Smiling, he scooped them up in his net. "I don't want it, it's a bulleye," he said of one. But he noted he could still get $2 for it at the pet shop.

While gentrifying East Williamsburg was no longer the pigeon hot spot that it used to be, Richie—like Panama and many of the flyers of color—could point to several coops in the vicinity where men he knew, and of his ethnic and age profile, kept pigeons. The myriad coops in Bushwick were close enough to provide him with some action, and Joey's pet shop was a short drive away. Thus, on a daily basis Richie hoped, and prepared for, an aerial clash; and he tended to have a short attention span when there was no action.

Dirty Work

Though Richie was gleeful over catching strays that afternoon, he put the event in perspective, chuckling, "Ninety percent of the time it's boring as hell!" He explained that, aside from the half hour or so per day that he got to fly the birds, and the far more infrequent opportunities to catch strays, there was "nothing to do but watch them putter around . . . and eat and shit." Richie was in fact in the process of getting rid of all of his "plain head birds" and replacing them with flights that had caps "just to try to mix things up and keep it interesting." As he scrubbed pigeon feces off some nest boxes, Richie—dripping with sarcasm—added, "Yep, it's a real exciting life!"

The amount of labor involved in keeping pigeons, a major part of what made it "boring," surprised me. Aerial encounters were peak experiences. But flyers' everyday routines more closely resembled a meticulous work schedule, especially as the hobby's declining popularity meant that most neighborhoods did not see enough "action" for flyers to catch many strays. This decline is reflected in language. The once popular Italian term for these tangles, "la guerra," and its English equivalent, the "pigeon wars," have faded from flyers' argot.[7]

My field recordings are filled with the sounds of hoes scraping feces, hoses spraying down rooftops and filling watering cans, and grunts from the men climbing ladders to the roof with 50-pound sacks of feed over their shoulders. I regularly found Carmine with a hammer in hand, replacing the rotted lip around the roof hatch, building a cage for a sick bird, or fixing his coop's screen door. Richie was making a "shanty" the first time I went on his roof. Joey constantly made repairs to his grandfather's antique coop while maintaining his own. The birds also required constant care. Just after the men got their coops winterized each year, and their birds finally stopped shedding scores of feathers that had to be vacuumed up daily, it was time to mate up the hens and cocks for breeding. Come springtime, the men tended to the babies, placed bands on their legs, vaccinated them, and changed the food and water more frequently. Then it was time to train a new cadre of pigeons how to fly together.

While the men did not particularly enjoy the grunt work, they did see the value in it and seldom whined about it. In an interview, Joey casually said, "I love the birds. I don't like the cleaning up, but that's part of anything." This line hints at an important way of seeing the world born out of experience. When Joey had to begin doing construction work after his son was born, I asked him how he felt about the transition from the pet shop to this backbreaking labor. He answered that he was pleased because it paid him $30 an hour. When I clarified that my question was whether he was concerned about the physical aspect of the job, he simply shrugged and said, 'It's not brain surgery, you know?' The bodily demands were routine for him. Joey's prior jobs had conditioned him to work with his hands, get dirty, and do what it takes to get the job done. This is important in both a *practical* and a *moral* sense. In a practical sense, doing gritty manual labor was a mundane experience for Joey, and had been since childhood, in a way that it is not for many middle-class, professional and service workers and their children. Carmine came from a time and place where one was not a man if he did not know how to fix his car, where boys lied about their age to sign up for the Second World War, and where they worked on the docks or in the factories when they came home from the war. When it came time to build his coop for his new home, Panama—as an electrician and manual laborer for over 30 years—did not think twice about constructing it himself.

Most if not all of the flyers I met had a blue-collar background. Only 3 of the 44 regulars I knew had college degrees, and only 2 of them had what would be considered "professional" jobs. None originated from wealthy or even solidly middle-class neighborhoods. A handful had union jobs that placed them in the middle class based on income, such as Panama the electrician, "Terry the sanitation guy," and "Tony the firefighter."[8] But their status and lifestyle were aligned with what they, and most sociologists, would consider working class.[9]

By virtue of their occupational and familial histories, the flyers appeared to have absorbed habits of body and mind that normalized and even valorized the manual labor of their jobs.[10] Carmine even claimed that he quit his well-paying job at the Navy Yard because his coworkers 'had every trick to get out of working,' and he 'couldn't stand it anymore.' He also said that, in his time as a loadmaster for Lufthansa airlines, 'I used to arrive an hour early just to get situated and see what had to be done for the day.' The stories the flyers told about their work experiences and those of their fathers were tinged with this ethos. And they extended this philosophy to their leisure pursuit of pigeon flying. These experiences also played a role in neutralizing the stigma that others might attach to such "dirty work,"[11] or dirty leisure. Most of the flyers were not caught up in trying to look like anything other than what they were. As people who might go to the pet shop or lounge at home with their respirators still around their necks and their clothes covered with paint, grease, or dirt from a long day's work, most of the men did not mind letting pigeon feces dry out on their clothes, even when they planned to go to the pet shop. In fact, these stains helped signal their status to knowledgeable others.

When the flyers worked on their coops and with their birds, their muscles did not ache, or the pain was simply dealt with. The hammer fit into the hand like a glove. Dirt did not offend the skin. The men's class or occupational backgrounds cannot satisfactorily explain why they kept pigeons. But their backgrounds did condition the satisfactions they obtained from pigeons by fostering a disposition that appreciated the physical aspects of the hobby.[12]

Work, sociologist Douglas Harper argues, has customarily been "the well from which the other components of the self are drawn." In this regard, the flyers' moral elevation of manual labor and self-discipline is typical of members of the working class.[13] Keeping pigeons entailed work, but, importantly, it did not have to involve as much work as the men put in. It appeared that flyers *elevated* caretaking to a job, and by investing so many hours of manpower, they came to invest their esteem in their coops and birds. Carlos Castro,[14] a Puerto Rican in his 50s with jet-black hair and a mustache, called other flyers his "colleagues." Possessing a master's degree in social work but hailing from a humble background, he declared, 'We're professionals. This isn't a bunch of boys up here with some cardboard boxes or crates.' Carlos described how he and his partner Ricky cleaned the coops every day and stuck to a strict schedule. He pointed out his medicine cabinet, filled with needles and natural remedies such as elderberry extract, and he encouraged me to examine the birds' feathers and eyes to see how well groomed they were, exclaiming, 'Fifty percent of it is work. We don't just come up here to watch the birds fly.'

Though Carlos's occupational status may have influenced his use of terms like "professionals," other flyers echoed his general sentiment. Panama called flyers his "associates," and he told me that it was his "job" to get his birds flying tight and to provide them with a clean and healthy environment. He explained, "I'm an electrician. As an electrician I gotta have a plan like two or three stages ahead of where I'm at. And so I take that to everything I do." Like Carmine, Panama signaled the moral meaning of putting in work by complaining about how "other guys" let their coops get filthy. Flyers treated their coops like little homes, decorating them, repainting them, and adding annexes. In doing so, they "inscribed their class-bound moral values into their physical surroundings."[15]

The flyers strived to maintain a sense of orderliness in their coops, the product of a regimented, worklike routine. This applied to their birds as well. None of them let the birds do whatever they wanted. Joey made sure his pigeons got one solid hour of flying time every day. Carmine's young ones went up every morning at dawn. Panama emphasized how he "punished" his birds with long flying sessions to develop a disciplined "power team." A strict enforcer, he sold pigeons to the pet shop if they came down early or landed on neighbors' rooftops. Sal Monaco, an Italian

American third-generation pigeon flyer in his mid-50s, mated up his hens and cocks on the same date every year and tracked the outcomes on a computer database. He let his birds take a bath only on Saturdays, and he flew his flights at the same time every day, for exactly a half hour. "Everything is done by conditioning," Sal explained to me one time on the roof as he hit the metal bars of his coop's open door to produce a ring. At this sound, his pigeons descended from the sky and marched one by one into the coop. Sal beamed, "It's like Pavlov's dogs. Very orderly—one, two, three." German sociologist Hans-Georg Soeffner found a similar "principle of affinity" among mid-20th-century miners who flew pigeons in Germany's Ruhr district, noting that the "training, discipline, and work ... required of the miner" were also the conditions that he imposed upon his birds.[16]

The flyers celebrated sweat and discipline and demeaned those who avoided it. This, they said, is what "real pigeon flyers" do. A man's worth lay in his deeds. Through their rooftop actions, the men performed and morally reaffirmed this classed and gendered discourse of dignity. Their esteem derived partly by aligning their leisure pursuit with this frame.

There is a vital difference between valuing manual labor as an ideal and investing one's esteem in his current job.[17] Most of the nonretired men held low-paying, low-skill jobs in which they had little authority. Such jobs, like loading a delivery truck, did not offer the same pay or respect from mainstream society as the union jobs that a lucky few flyers held. In this context, pigeon flying was akin to a self-fulfilling, skilled, and chosen job where flyers could base their esteem on the fruits of their labor. Caring for pigeons takes expertise. Because of a lifetime of experience, the men had a competent knowledge of operant conditioning and genetics that enabled them to train the birds and predict the outcome of breeding. Keeping pigeons offered flyers the satisfaction of a stable routine with obligations that must be met, yet had enough unpredictability to keep it interesting. Carmine and Frankie often woke up with their whole day mapped out, including well-earned "breaks" for coffee, and on any given day, things could be spiced up by a hawk attack or by the occasional capture of a stray pigeon. Rather than sitting idle or passively having their life structured by television, through pigeons these retired men provided their own structure and narrative to their life—the birds and coops required their constant labor and attention. Interestingly, the work of keeping pigeons (e.g., feeding, bathing, and raising them) also has clear parallels to the kind of domestic "care work" that is traditionally coded as feminine. Though none of the flyers framed it this way, pigeons seemed to provide an opportunity for the men to perform care work without it posing a threat to their masculinity.

Nature and the Social Self

The flyers' intimate contact with pigeons fostered a degree of emotional attachment. Every year, for instance, Carmine found joy in breeding a new stock of young ones and training them to route. These interactions offered him opportunities to nurture and to even try to imagine the animals' point of view. I watched as Carmine spent most of one spring day trying to lure back to the coop a lost young one that had flown for the first time. Finally, at sunset, it returned. Carmine beamed and said to me, 'I know my birds! I told you he'd be back.' Holding it, he said he could feel that it already lost weight. He spoke gently to it, saying, 'You learned a valuable lesson,' as he isolated it to ensure that it could eat unmolested. While losing birds to disease and the hawk was unavoidable, the cycle of life brought new chicks to replace them. When I visited him on a rainy April day, he proudly set some awkward babies down on the bench, saying, 'Look at this beautiful Isabella! And look at that nice silver dun! Their caps are so beautiful already, imagine what they're gonna look like in a few weeks!' He belly-laughed, then said softly to the chicks, 'I know you're thinking, what is this wet stuff hitting me? Where am I?'

The eminent biologist and naturalist Edward O. Wilson sees relationships like those that Carmine had with his pigeons as a manifestation of what he famously calls *biophilia*, defined as "a human *need* for deep and intimate association with the natural environment, particularly its living biota." The desire to get close to and understand life processes, it is contended, is an adaptive hereditary response "fired in the crucible of evolutionary development" that also, over time, has become encoded in language and culture.[18] Although proponents of the "biophilia hypothesis" concede that this innate drive can be conditioned or stymied by social forces, they maintain that our responses to nonhumans are "biased in certain directions by our evolutionary history."[19] It is biophilia, they argue, that is at the root of our awe upon encountering towering sequoias, our desire to draw pets into our families, our exuberant response to monkeys in the zoo, and the ubiquity of animals in language, myths, and fairy tales. And it is biophilia that influences so many people to mourn the decline of biodiversity and dedicate their lives to protecting other species.

Humans are wired to consider other animals as kin, Wilson stresses, by virtue of sharing a genetic code. Biophilia proponents point to many premodern cultures' adoption of nonhumans as sacred totems as evidence of the pervasiveness of this feeling of kinship before the scientific and industrial revolutions. By rationalizing and commodifying nature, and severing our intimate link to it, these societal forces are said to blunt the biophilia impulse. Nonetheless, ecologist Stephen Kellert argues, humans' search for a "fulfilling existence" is still "intimately dependent upon our relationship to nature." And even being in "right relation with but one corner of creation" can foster a sense of connection to—and shared destiny with—the environment.[20]

Sociologist James Gibson sees in biophilia a sociobiological complement to his cultural thesis on the "re-enchantment" of nature. Gibson argues that modernity severed the "traditional unity between humans and the rest of creation," but that more and more people today, driven by a need to transcend the ills of society and find fuller meaning in their lives, are "seeking a new kinship" with "animals and landscapes."[21] Echoing Kellert, Gibson claims that personal relationships or "powerful encounters" with a single animal or plant can foster a sense of primordial affinity with nature in general. He argues, for instance, that an old oak tree in California that was to be cut down to make room for a new road became "far more than a tree" to the local community; "it had become a symbol of all other trees, animal life, and open spaces lost to development."[22] And he interprets many New Yorkers' celebration of the red-tailed hawk Pale Male, who took up residence on a Fifth Avenue edifice, as a consecration of nature and the expression of a desire to escape the concrete jungle.

The biophilia and "culture of enchantment" theses both frame people's desire for relationships with other species as primarily rooted in an internal need to connect with nature, and they both argue that such associations enable people to transcend social life. Yet while this perspective may help illuminate one way that people experience encounters with nonhumans, sociologist Leslie Irvine points out that it neglects "the potential and various meanings that individuals give to their relationships with nature and other animals."[23] Attempting to reduce all cross-species relations to biophilia leads to unsatisfactory explanations. To wit: Kellert argues that "even the tendency to avoid, reject, and, at times, destroy elements of the natural world can be viewed as an extension of an innate need to relate deeply . . . with the vast spectrum of life."[24]

Both the biophilia and "culture of enchantment" theses overlook the extent to which *close relations with nonhumans may be catalyzed or sustained by* social impulses *and enable a sense of connection to the* social world. Most flyers were fascinated by pigeon biology, genetics, reproduction, the homing instinct, and so on. Yet when I asked them if they felt an affinity to nature, I was usually met with a blank look followed by a simple "no" or "not really." I saw no evidence that pigeon keeping was part of, or led to, a more general connection to nonhumans. A few of the men had dogs, but that was it. Their interest did not even extend to other pigeons. They called feral pigeons "street rats," and most were loyal to a single breed—often flights, sometimes tiplets—and considered other breeds "garbage." Even representatives of the flyers' favorite breed were often considered garbage if they did not hew to an imaginary ideal.

The men were thus attached to their birds not primarily because they were ambassadors of the wild but because they were products of the men's own hands. In the book *Dominance and Affection*, geographer Yi-Fu Tuan argues that part of the pleasure of domesticating plants and animals comes from overcoming nature's recalcitrance. Manmade breeds are displays of "human ingenuity and power expanding into and subduing nature."[25] Regarding pigeons, author and fancier Stephen Bodio writes, "You can *be* natural selection, with the power to change the looks and even habits of an animal . . . [and] steal your neighbor's flocks . . . even if you are tied to a patch of city rooftop and a demeaning job."[26] Yet while domestication entails domination, Tuan also emphasizes the pleasure derived from having other species respond to human care. While nature writers like Gibson view dominion as antithetical to close and respectful relations with other species, Tuan concedes that human dominion over nonhumans can at times *enable* genuine affection for them. As people's relationships with pets reveal, intimacy thrives on dependency. Though the flyers seldom doted over their pigeons like pets, it was only through crafting and taming these birds that the men were able to make pigeons into objects of their affection.

The pigeons became an extension of the flyer's self. In caring for and directing the pigeons, flyers underwent a sort of "metaphysical merger" with their stock.[27] While particular birds came and went, the stock personified a living history of the flyer's efforts, a unique and continuous bloodline of his invention. And in carving out spaces for their birds, training the pigeons to hover over the streets, capturing others' birds, and evading the hawk, the men literally expanded their presence and influence into the surrounding airspace. While the men struggled to find words for this magical corporeal experience, Sal likened it to "flying a live kite." Whether the flyers lived in "dead zones" or in areas with a lot of "action," their animal practices were a significant way that their urban environment became relevant, knowable, and malleable to them.

Though every flyer had personal preferences (e.g., flight or tiplet, cap or plain head, solid color or teager pattern), their tastes were hardly idiosyncratic. While Tuan emphasizes the *psychological* satisfactions of breeding, the flyers were, to use a phrase from sociologist Jeffrey Nash, "breeding for *social* meaning."[28] This is why no other variety of pigeon mattered, and why pigeons did not seem to make the flyers feel more connected to nature. The pleasures of breeding and training came from creating specimens and flight patterns that the men had been socialized to value through their peer group. These standards were codified in a place-based folk knowledge passed down through generations of flyers. We can never know if something like biophilia played a role in attracting the flyers to pigeons as children, but what *is* clear is that their social world guided these relationships and endowed them with meaning. And, while it is unclear how innate human drives may have shaped the men's animal practices, it *is* apparent how these cross-species encounters were patterned by social forces like class and gender.

In the book *Mind, Self, and Society*, the influential social psychologist George Herbert Mead argued that one's sense of self comes from "taking the role of the other," imaginatively looking upon oneself as others do. The self is *social* because it is arises through interaction with other people. We respond not to others' actions directly but to our mental assessment of how they will interpret our response. And it is "due to the individual's ability to take the attitudes of these others . . . that he gets self-consciousness." Interestingly, Mead suggested in a footnote that nonhumans could play a part in the constitution of the social self. This is possible because one can still respond to "objects socially or in a social fashion"—invoking the attitudes of others—and in the process become "conscious of himself as an object or individual."[29]

Though Mead's 75-year-old theory remains a cornerstone of sociological understandings of the self, his footnote on the role that nonhumans can play in constructing it continues to be an afterthought. Sociologists Clinton Sanders and Leslie Irvine, however, point to some of the ways that relationships with pets shape the self and social relations. They show that people routinely assume their pet's point of view and act as if their pet can do the same. When humans interact with pets in this way, Irvine writes, they incorporate their imagination of their pet's perspective into their sense of who they are as a person (e.g., loving, loyal, fun). Sanders illustrates the wider social ramifications of these dyadic relationships by examining how others treat guardians and their pets as a "couple identity" in public. When the dog is well behaved, it reflects well on the owner; when the dog misbehaves, however, human guardians quickly proffer excuses and engage in "repair work" because they sense that other people's opinion of them will be sullied by their dog's behavior. In this way, dogs influence the social self and interpersonal encounters.[30]

In the case of the rooftop flyers, the men's relationships with their pigeons clearly organized their social relationships and formed a core part of their social selves. The pigeons anchored the flyers in their neighborhoods, becoming a linchpin between the individual and his community, the past and the present. For the ethnic white old-timers, pigeons were their rudder in a sea of change. For the younger men of color, through the birds, they interacted above the streets by coordinating their flying times and catching each other's birds. And, as shown in chapter 5, pigeons brought these racially and age-diverse men together as a peer group. Even when the flyers were alone on their rooftops, their peer group lingered in their consciousness. It was through taking the role of the other flyers that the men determined how to breed, train, and appreciate their birds. Reciprocally, the flyers' social selves were honed in their interactions with pigeons because they knew that the status other flyers conferred on them was based on their birds' appearance and performance. And, finally, as men who morally valued manual labor, pigeon flying became a means of performing and affirming their working-class persona.

While sociologists occasionally acknowledge that interpretations of animals and nature "reflect," or are "grounded in," particular social contexts, the field typically ignores the ways that nonhumans can actually organize the social realm.[31] For the rooftop flyers, their pigeons dictated the men's daily routines, their experience of their neighborhoods, and their social relationships. They embedded the flyers in a distinct social world and guided—not just mirrored—their sense of who they are. There is also an important lesson here for environmental scholars who assume that close relationships with animals are fulfilling mostly because they fulfill some sort of need to connect with nature and transcend social life. Close relations with other species may also be compelled by socially derived inclinations and may augment connections to society. I see no reason to view such relations as somehow less "meaningful," "authentic," or "pure" than the mystical animal encounters often highlighted by biophilia advocates. As I will show in the next chapter, these two modes of experience are not nearly as distinct as they seem.

As I sat with Carmine on his rooftop one quiet morning, lazily watching his pigeons chase after the grains of corn he was tossing at them, I broke the silence by asking Carmine what was left for him after all of his children and ethnic peers moved away or died. He chuckled for a moment before soberly replying, 'All I got is my birds. I got nothing else.' Carmine's enduring passion for breeding and training his birds endowed his life with purpose, and his pigeon loft acted as an anchor and a compass as he navigated his ever-changing neighborhood ecology.

FOUR

The Turkish Pigeon Caretakers of Berlin

Primordial Ties in a Migrant Community

ON THE NORTHERN border of the Turkish neighborhood of Kreuzberg, on a vacant piece of land just east of Michael-kirch-platz and about one meter from where the Berlin Wall once stood, sat a pigeon coop. I had come to Berlin in the summer of 2005 to attend a seminar, and I stumbled upon the coop while wandering the streets in my free time. Intrigued, I returned the next day with a German-born Turkish sociologist whom I had befriended, Beyhan Yildirim. Beyhan translated between Turkish and English as we talked to Ahmet Çabakçor, the main "caretaker" (his word) of the coop's domesticated pigeons. A Turkish migrant and television factory worker from the city of Sivas, Ahmet was 45 years old at the time and had a trimmed mustache and cropped salt-and-pepper hair. He was a warm, intelligent, intensely nationalistic fellow.

Ahmet's pigeons were called Turkish tumblers, and he made the origin of the breed's name immediately apparent to us by tossing a pigeon upward like a ball with an underhand scoop. After it ascended several dozen yards, the bird repeatedly performed backward flips in midflight until it returned to the ground. As he leisurely lobbed more birds, Ahmet emphasized that Anatolian sultans, soldiers, and peasants originally bred tumblers, and that specific regions in Turkey were even known for their own distinct varieties. "The reason why I am interested in these pigeons," he explained, "has to do with my ethnic origin. In Turkey, these pigeons have an old tradition dating back for centuries. My family used to keep

---

Research for this chapter was conducted in collaboration with Beyhan Yildirim.

them for fertilizer. They took care of them on the roof; everyone had pigeons in Sivas." Yet when he migrated to Berlin, he "was more focused on the labor market" and did not keep pigeons. "But I was always thinking of pigeons," he recounted with a smile. "If you grew up with pigeons, if it is in your culture, then it is in your blood. You can never forget this as a child, and I had this experience, I had this passion." After 12 years in Berlin, he finally acquired this land and some of his own tumblers. Ahmet remarked that he did not know at first that keeping tumblers in Berlin "could help with social relations," but through his birds he formed close ties with other Turkish men who shared his passion. He added that keeping pigeons was "a much healthier activity than just hanging out at the teahouses, drinking"—a trap he believed many of his coethnic peers fell into.

Though I spent only two days with Ahmet before I had to go back to New York, I returned to Berlin for three weeks in September. Beyhan and I visited eight coops and observed and interviewed 26 caretakers, all but one Turkish, during that time; but we spent the most hours observing the interactions at Ahmet's Kreuzberg coop because it seemed to be the most central gathering place. The space, which Ahmet shared with five others, usually had at least three to six men sitting on chairs in the yard or working with the birds from morning until twilight.

The intense homesickness and nostalgia that often accompany migration partly originate in the severe rupture between the culture and lifestyle of home versus the host country. For Ahmet and the other pigeon caretakers we came to know, an especially significant void they said they felt in their transplanted lives was the lack of a direct connection to nature, especially through relationships with animals. Yet, hidden in urban interstices, such as fallow areas where the Berlin Wall stood or behind an abandoned factory, these Turkish men built places where they could feel immersed in nature through their close relationships with pigeons.

The caretakers narrated their bond with tumblers as "in the blood" and, unlike the New York flyers, said they felt closer to nature through interacting with pigeons. Yet I found that the men's primordial attachment to tumblers was primarily rooted not in the belief that the birds provided "unmediated exposure to [the] ultimate reality of nature,"[1] but rather in the feeling that "Turkish blood" coursed through both human and avian veins. The tumbler's mystical aura emanated from its connection to the Turkish homeland, and the men considered their attraction to nature to be a Turkish cultural trait. The men's reverence for tumblers, and for

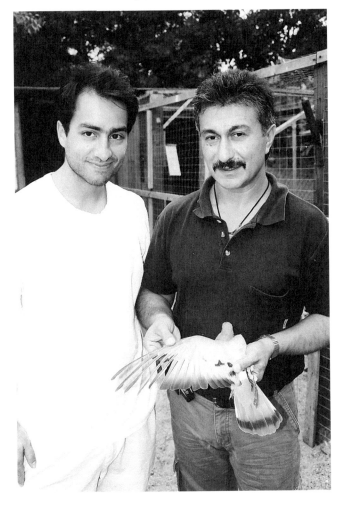

Ahmet inspects a bird with Beyhan.

nature in general, was a means of consecrating *nation*. In this way, caring for and affiliating with pigeons was a performance of ethnicity,[2] and the immigrants' coops enabled them to eke out a Turkish social space in the midst of a foreign city. The caretakers indicated that even the quest for a sense of "kinship" with nature—which environmental scholars usually depict as driven by a yearning to "transcend" social life and access "a numinous world beyond our own"[3]—can be inspired by, and ground us in, our social self. Moreover, the men revealed how human-animal relations can play a significant role in the formation and maintenance of ethnic identity, even helping immigrants mitigate the psychological and cultural trauma of being uprooted.

The Caretakers, Their Coops, and Their Birds

Germany is home to the largest population of Turks outside of Turkey, numbering roughly 2.5 million. Many found their way into West Germany and West Berlin as "guest workers" (*Gastarbeiter*) in the 1960s and 1970s, promised decent wages and a better life in exchange for providing the cheap, relatively unskilled labor needed for the manufacturing economy. Scores of these migrants settled in Germany and created ethnic hubs in urban centers such as Berlin, which today hosts at least 130,000 people with Turkish citizenship and 50,000 Germans of Turkish descent. They stayed put even though the manufacturing jobs dried up and they faced persistent discrimination. But while Turks have traditionally been forcibly marginalized to ethnic colonies, today some prefer to live almost exclusively within Turkish districts and have minimal day-to-day contact with German society.[4] A large number of Turks in Berlin are unemployed,[5] and most of the unemployed do not have a high school degree.

Kreuzberg, the neighborhood in Berlin where Ahmet's coop was and where many of the caretakers resided, is one of the best-known "Turkish" enclaves in Germany. But this label elides its heterogeneity. Centrally located (but on the edge of West Berlin until the wall came down), it has long been a haven for many migrant groups and German bohemians. It is not a blighted or isolated ghetto, and other Turkish neighborhoods in other cities have higher concentrations of Turks. Wedding, another Turkish enclave three miles northwest of Kreuzberg where several caretakers lived (and where I stayed with Beyhan), resembles Kreuzberg in many ways.

The pigeon caretakers generally matched the demographics described above. Due to waves of prior migration that created vibrant Turkish enclaves, they were not socially isolated or without families or kin. Most came to Germany with their parents, who were guest workers.[6] With two exceptions, all were married with children. None had a college degree. In 2005, the mean age of these men was 39; the youngest was 19 and the oldest was 65. Two of the men were unemployed; three others were only marginally employed. Four of them worked in an *imbiss* (a small restaurant specializing in kebabs), a common occupation for Turkish men,[7] while most of the others worked in unskilled light manufacturing jobs. One was a car mechanic, and only two had positions of any authority—one as a construction supervisor and another as a building superintendent. The group's rate of employment was above average for Turks in Germany. The outlier in the group was Sinan Samil San, a highly ranked professional heavyweight boxer who lived in the suburbs with his wife and child and who knew only a few of the caretakers. In addition, a young man from Kosovo had Ahmet's blessing to visit the coop and fly the pigeons.

Finding a plot for a coop required negotiation and ingenuity, and the coops shared the quality of being hidden. Only two of the men had a coop in their own yard. Two men rented a fallow plot of land, one from another Turk and one from the municipal government. One even "squatted" a piece of land downtown where the Berlin Wall stood. Ahmet's coop sat on an empty lot that once faced the western wall (Ali, a frequent visitor to the coop, remarked that the pigeons that flew over the wall into East Berlin were never seen again). Ahmet said he acquired the space from a friend who retired to Turkey after being one of the first people to introduce tumblers to Berlin in the 1980s. Ahmet seemed to have inherited the man's role as the perpetuator of a small but growing number of pigeon caretakers in Berlin.

In the midst of the densely packed and chaotic, yet colorful and amiable, Turkish enclave of Kreuzberg, with teahouses, grocers, and kebob shops dominating the storefronts, Ahmet's coop offered a trace of bucolic tranquility. Carved out of an acre and a half of undeveloped land that abutted two streets and two buildings, the 25-by-12-foot plot was encircled by a low wooden fence and some shrubs. The surrounding trees and the coos of the pigeons buffered the traffic noise, and the dull, musty smell of pigeon feces gave the environs an earthy scent. Two simple wooden structures with tiny windows and doors served as bookends of the plot, joined together by an "aviary" that was made of two-by-fours and chicken wire. In front of the aviary was a small dirt yard. One of the wooden structures housed the pigeons, while the other acted as a sort of clubhouse, with old photographs of show pigeons, a stove to brew tea, and a bench to enable socializing in the cold winter months.

The type of tumblers that the men kept were originally bred from the common rock pigeon centuries ago in central Asia, especially in the region of present-day Turkey. Tumblers look similar to common pigeons but usually have feathers on their feet (which fanciers call boots) and have been bred to tumble. Darwin speculated that fanciers selected and exaggerated a spastic, seizurelike inherited defect until a tumbler was eventually produced.[8] Though training can enhance their tumbling ability, the birds seem to have an innate and partially involuntary proclivity to tumble. When released they will sometimes tumble repeatedly while falling toward the ground and land on their backs. The men's birds were either tumblers actually brought from Turkey or birds bred in Germany out of tumblers originating in Turkey.[9]

Of Nature and Nation

During a visit to the coop of Israfil Durkut, a 41-year-old construction manager from Sivas with curly gray hair who moonlighted as a wedding singer, he pointed out his favorite pigeon. "It's a blue one. It's my central nerve!" He held up a shiny gray tumbler with feathered feet and tossed her skyward. "Wow, it's tumbling a lot," I remarked through Beyhan. "Is that why it's your favorite?" Grinning, Israfil responded while following her flight path with his eyes, "The reason is that I can identify myself with her." Seeing my quizzical look, he laughed as he continued, "I have known pigeons since I was four. I had my own, and I have seen many pigeons from other guys, but this one is incomparable. I have never seen such a good tumbler. She shows her characteristics and features, her personality. I wouldn't sell her for €20,000." I asked Israfil how he got into pigeons, and he explained that his father cared for them. "As a child, pigeons were my closest friends. When I was a little kid I was especially interested in the 'voices' of the pigeons—their coos. I was so in touch with them, they weren't afraid of me. I was interested in animals in general, but especially pigeons." Israfil said he migrated to Germany with his parents in 1979 at the age of 14, and that ever since he first arrived, he missed his pigeons.

"The first place I worked in Germany was in a garden, a courtyard. My employer had homing pigeons. I stole some from him—babies. I tried to take care of them, but they returned to the courtyard." Dejected, Israfil said he would lie in bed at night and cry if he heard pigeons cooing on his windowsill, remembering the pleasures of "running on the roof and through the streets to catch my own pigeons" when he was "a little kid in the village." "I later took street pigeons and made them pets," he recounted, "but it didn't work out either." For a long time, Israfil did not have his own pigeons but befriended and spent time with some of the other caretakers, until he finally found this space. Israfil had taken over a fallow swath of land in downtown Mitte, within view of the Berlin Wall museum and with an original piece of the eastern wall as one of his boundaries. Largely out of the view of passersby behind a flea market, he kept his pigeons in a white storage shed and erected makeshift wooden fences around this piece of earth so that no one could see inside. He now spent three hours a day after work here. "If I am with the pigeons, I don't think of anything else. I am just focused on the pigeons."

Every caretaker narrated a childhood connection to pigeons that was initiated by a male relative in his homeland. It was common for the men to claim that the remembrances and connections to childhood that resulted from taking care of pigeons in Berlin were a significant motivating factor to rekindle the practice. As Ahmet informed Beyhan and me on our first visit, "If you grew up with pigeons, if it is in your culture, then it is in your blood. You can never forget this as a child." Their attachments to the birds in Berlin mirrored the men's childhood ties. Ahmet Çabakçor's friends, Ahmet Dede and Mehmet Ünver, both told me that their childhood fascination with pigeons even got in the way of school. Orhan, a balding 50-year-old car mechanic from Ankara with short gray hair and stubble on his chin who was a smiling and constant joker, concurred: "When I was a child, if I lost a pigeon I would see it in my dreams. I couldn't sleep. I would wake up the next morning before sunrise to look for it." If it was a school day, Orhan recalled that he would skip school to find the pigeon.

As Orhan kissed a pigeon before releasing it into the air, he told a similar tale of disruption and yearning for pigeons. Though Orhan said he had been in Germany since 1979, "until last year I didn't know that people took care of pigeons in Berlin. Last year I started; I had 15 pigeons. Now I have 60, and I gave 20 to my friends." Through finding a network of caretakers, Orhan realized it was possible, and gained the motivation, to construct a pigeon coop on the gravel lot in front of his auto garage. Since then, his friend Asim Çím, a gentle, bearded 40-year-old from Konya (in central Turkey) who migrated to Berlin at the age of nine, had joined him. "I skipped school because of them. One time I looked for one of my good tumblers and wound up looking for three days, missing school," Asim recounted with a laugh while applying ointment to a cut on a pigeon's wing. "I always had them on my mind, but I couldn't find the space. If you are involved in taking care of them since childhood, then you can't forget them."

The love for pigeons appeared to be linked in their minds to positive experiences that the men had with these birds in the past. Such experiences opened them up to, and oriented them toward, meaningful interactions with tumblers in Berlin. Yet the intergenerational transmission of this animal practice hints that the men experienced the return to pigeon caretaking as more than just a chance to enjoy the company of these animals. The men interpreted the attachment as an almost magnetic force, struggling to find the words to express the feeling of the tie. As we did with all of the caretakers, Beyhan and I pressed Asim to talk about the draw of pigeons. He answered, "It is definitely a disease, a serious one!" Such hyperbole was commonplace: all but one of the caretakers interchangeably referred to caretaking as a disease or "in the blood."

Abit and Abdi Yasaroglu, two thin and quiet but affable brothers in their mid-30s with dark skin and hair, told us they were born in Germany but sent to live with family in Istanbul at the ages of six and seven. Abit claimed, "My uncle infected me" with the desire to care for pigeons, and on a different day Abdi stated, "I stopped going to school because of the pigeons. I repeated one class three times, just because of pigeons! This hobby can lead to a disease." While Abdi was fond of this animal practice, he said he had reservations about his children becoming "infected." "The activity reminds me of my childhood, and of my native country. But I don't want to infect my children with this disease [laughs]. At this stage, it's not good for them, because of school." Indeed, the brothers appeared to be unable to pull themselves away from their birds. On the day we met them, their mood was sour because one of their pigeons was missing. They sat in chairs, eyes continually upward, cupping a bird in each hand. They frenetically scanned the sky, occasionally tossing a pigeon in hopes of luring the lost one back. They also waved a pole with a blue plastic flag—which, like the New York flyers, they used for training—to rustle the birds in the coop in the hopes that the sound of their flapping wings would be heard by the wayward bird. They whistled and shook a can of food, they canceled their afternoon plans, and they waited—for hours. They talked about how this was a common occurrence, and if all else failed, they would stay until sunset and hope that hunger would drive the pigeon home. Later, as nightfall descended, they put their birds away and vowed to come back first thing in the morning to continue the search. Just as we were departing, Abit shouted, "Shhhh!" The lost bird had come back and cautiously landed atop the aviary. Abdi slowly crept to open the coop, wearing a fierce look of determination. As the rest of the flock emptied back into the aviary and we held our breath, they tossed some grain to coax the skittish straggler in. Finally, it dropped through the trapdoor, and the brothers let out triumphant yells. It seemed as if a great burden had been lifted from their shoulders, and they happily departed for their homes.

Many of the men showed devotion akin to Abit's and Abdi's. Suat, a

Abdi trains the birds while Abit watches. Photo by Oliver Hartung.

25-year-old man with a boyish face and chin-length blonde hair who grew up in Ankara, revealed how addictive tumblers could become. His coop was a small white trailer sandwiched between a cement canal and a chain-link fence behind an old factory. Unemployed and single, he was spending upward of 10–12 hours a day at his coop. Suat meticulously trained his pigeons, documented their tumbles, and groomed them. When we asked him what the attraction was, he ashamedly replied, "It's an obsession, a disease. It's like if someone is an alcoholic [laughing]. It's like tuberculosis. You forget your children, your job, your family, you forget everything once you are involved in this." The next day, Orhan told us, "It's a disease, once you begin with this you never stop until the end of your life. There are people who are 90 years old. They cannot walk. They cannot sit. They still take care of pigeons." And Sinan, the square-jawed and chiseled boxer, opined, "You may think I'm stupid for keeping pigeons. You need to have this desire by birth." He pointed at his heart for emphasis.

This rhetoric of disease served as their folk explanation for why the men returned to keeping pigeons, some after a prolonged absence. Timur Cesur, a 39-year-old man from a small town 120 miles west of Istanbul with thinning slicked-back hair, kept his pigeons on the roof of an Internet café he worked at near Tegel airport. We visited him on a late afternoon and watched him toss several pigeons into the sky. As they flew toward the clouds, tumbling and then circling among the low-flying planes, he lit a cigarette and began: "I got in touch with pigeons through my uncle and my grandfather. I was seven. I had them until I came here in 1992. Until 1995 I didn't have them, and in 1995 I brought four here illegally." Describing the attraction, he continued, "It's kind of a disease—as soon as you have the ability to keep them again, you start." We asked him to describe this disease further: "The love is instinctive. It exists in everyone. Everyone has this instinctively, and you never know when this instinct will come out."

Though the amount of time varied, each caretaker told a story of being without pigeons for several years until he met other caretakers. For example, Abit and Abdi began keeping pigeons again after they met Ahmet and he donated some birds and a space to them (as he did for Mehmet, Ahmet Dede, and Suat). Keeping pigeons was a social activity, initiated and sustained by a group. What were the caretakers expressing, then, by claiming that keeping pigeons was "in your blood" or a "disease"? Suat's description of caretaking as akin to alcoholism is perhaps most apt for explaining the "career" of these men: seduced by pigeons at a young age through relatives, once bitten they felt unable to resist the temptation when an opportunity was presented.

By claiming that pigeon keeping was a disease or in one's blood, the caretakers expressed a connection to the animals that was intimately bound up in their feelings of rootedness in their homeland, refracted through their nostalgic remembrances of their childhoods and an apparent intense homesickness. They also had warrants for thinking that it was literally in the blood because it was passed down to them by their fathers, grandfathers, and uncles and was (in their eyes) such a part of Turkish male culture. In fact, the maiden name of one caretaker's mother was Kuşçu, which means "caretaker of birds." Ali Belli, a 65-year-old window framer from Amasya who sported a gold pigeon medallion around his neck, proudly emphasized that his grandfather chose this name when Turks were first required to have surnames in 1934. Claiming caretaking was in the blood also appeared to be a way of finding expression for the feeling of the "ineffable, and at times overpowering, coerciveness"[10] of the bond, as many of the men truly seemed to think that in the end, like it or not, they were compelled to care for pigeons.

For the men, continuing the tradition in Berlin also facilitated the forging of a quasi-kinship with the pigeons. While almost all of the men emphasized which—and how many—of their relatives kept pigeons in

Turkey, they were also interested only in pigeons that originated from Turkey. Some even told us that they smuggled anesthetized birds into Germany in their suitcases on airplanes. The pigeons were admired because their blood traced back to Turkey, just as the men's blood did. When Timur first showed us his coop, he insisted, "One hundred percent of [these] tumblers are from Turkey . . . Their historic homeland is central Asia. The tumbler is specifically Turkish." Orhan called his tumblers his children. Ahmet put it this way: "The Ottomans are in our genes, [and] there were princes who took care of this bird"; and Abdi stressed: "This activity reminds me of my childhood, and of my native country. All the birds are from Turkey." Asim told me, "The pigeons in Turkey are famous, [and] this reminds me of Turkey. Our ancestors kept tumblers." I also noted that some men especially prized tumblers from the same city as them, such as Ahmet's attraction to pigeons from Sivas.

Specific references to regions were not always on the caretakers' lips; often, they emphasized more general narratives about connections to nature. When discussing their love of nature, however, it was common for the caretakers to frame it as a feature of Turkish culture. In one of our meetings, Timur appeared anxious when a hawk chased his five airborne pigeons. His birds scattered, and he lost sight of one of them. His eyes darted about the sky, and he rushed to let his other pigeons out as he tossed seed in hopes of luring the lost bird home. While doing this, he told us, "The interaction [with people] is very nice and very good. But the most important is love toward the animal—to love and to understand the birds. In the end, it's about loving the animal." I asked Timur to elaborate, and he gave a paradoxical reply:

> My understanding, my philosophy is, you need to love and understand animals in order to love humans . . . For example, I am very angry at the hawk, but on the other hand I can understand the predator. It needs the pigeons in order to survive, and I have respect also for the predators. There is a saying "The more I learned about animals, the more I understood them, the more I started to hate human beings." The reason is that animals kill animals to survive but humans kill other humans . . . for their own interests.

I asked Timur how keeping pigeons was different here than it might have been in the small village he was from in Turkey. I saw buildings and concrete in every direction from his roof, and planes flew very low overhead. Yet he answered, "Being in an urban center doesn't matter. I just love nature. Even if we are in the city, we create our own natural environment. It's about being in nature. I can meet my friends and we stay outside. We all love nature."

Just then, the lost pigeon returned. Timur smiled, tossed some seed, and attracted the pigeon so that it landed on his hand. He continued, "I am only indoors when I have to be. When I am not at home or at work, I am with the pigeons. This means for me that I am outside and in nature. I do not consider this environment—I perceive it as nature, because I am focused on the pigeons. I don't see the buildings."

"Animals have a deep history in Turkish culture," Timur added, now giving his appreciation for nature a cultural and historical skin. He told me of the reverence of Turks for horses, and how wolves are held in high regard because they are independent and rugged—just like the Turks. Returning to pigeons, Timur said, "You cannot torture or eat these birds. The reason is they are protected by the prophet Mohammed." I asked him to say more, and he told a story that I would hear again on three other occasions:

> When Mohammed tried to spread Islam, he was forced to leave Mecca . . . So Mohammed was running away and the rulers of Mecca were chasing him. Mohammed was hiding in a hole in a mountainous area. Where the hole was, the spiders made a web over the hole; in front of the web, the pigeons made a nest. The people who were chasing Mohammed saw the pigeons in front of the web and said a pigeon would never leave his eggs here if there were human beings around. So they would lose time if they went into the hole, they thought.

Ahmet Çabakçor also spoke of a general love of animals and of the human-animal bond as an important historical reality of Turkey. He saw his activity in line with this narrative: "My city is also famous for horses and dogs. My grandfather cared for horses from central Asia. I continue this tradition today, even in Berlin. It comes from love of the bird, love of nature." Asim framed it similarly, telling me, "The pleasure comes from love for the birds. All Turks love animals. We have dogs, cats, horses, tumblers." Even the heavyweight boxer Sinan Samil San, 31 years old and from the city of Kars in northeast Turkey, kissed his pigeons and proudly showed me a baby that he said survived only because he chewed up the seed and hand-fed it. He talked "baby talk" to his pigeons and discussed feeling a connection to all animals. Indeed, his small yard was a menagerie—pigeons, a dog, cats, chickens—and he told me he also had horses in Ankara.

The ethnic dimension was not equally salient among all caretakers. For example, the men were split on whether they knew the history of pigeon caretaking in Turkey, or if it mattered. Ahmet was very knowledgeable about this history and firmly planted his practices within this narrative. Timur demonstrated this connection when he remarked, "Turks have been interacting with pigeons . . . [for] 5,000 years." Asim read books on the history and said, "Our ancestors kept tumblers. I don't want this to die out . . . It's very important to keep this Turkish practice alive." Yet some of the caretakers seemed to have little grasp of this history. Mehmet awkwardly admitted, "Ahmet knows [the history] from A to Z; I know only A to C." Others downplayed the significance of pigeons to Turkish history. Turan Demir, a small 50-year-old man from Sivas with gray hair and a mustache who shared a garden coop with his friend Hasan in the district of Wedding (but has since returned to Turkey), stated, "The history is not important . . . I cannot be proud that our grandfathers or Ottomans took care of pigeons. It would be important if they had invented something. I would be proud of that." Israfil added, "I don't care [about history]. I just love to hang out with them." Israfil may not have cared about textbook history, but he, like the others, repeatedly emphasized his love for nature and how the birds connected him to his homeland and his childhood. While the concern for *labeling* caretaking as a "cultural" activity might vary, these connections, and the belief that Turks have an innate love of nature, appeared to be a constant that promoted a sense of cultural continuity between the present and their idealized past.

Social Ties and Place

Though the men emphasized personal biography in explaining the draw of pigeons, their animal practices were embedded in the kind of social networks that are the foundation of community. Caretaking was also embedded within certain places to which the men formed sentimental, reverential attachments. The men's communal relationships created through pigeon caretaking, and the physical environments where these interactions occurred, played a key role in reinforcing their cultural interpretation of this animal practice. While each man framed connecting to nature as part of his Turkish identity, the caretakers collectively put such understandings to social use to condemn those who frequent urban teahouses and to define their social group—and the pigeon coop—as morally superior.

All the caretakers except Sinan, the boxer, proclaimed that caring for pigeons "keeps me out of trouble." Even the men who said that caring for pigeons disrupted their schoolwork believed that it kept them out of trouble in later life. But their moral denigration of men who visit Turkish teahouses added a layer to this narrative that had significant bearing on their ethnic identity. When Ahmet first mentioned that caring for pigeons kept him away from the teahouses (*kahvehane*), I thought little of it. Yet aversion to the teahouses almost seemed like a puritanical obsession among the caretakers. I talked to each man separately, and aside from Ekrem (who was from Kosovo, not Turkey) and Sinan, each said that it kept them away from teahouses. What exactly was at stake when the men contrasted the pigeon coop to the teahouse?

Teahouses, or coffeehouses, are an important social institution in the lives of urban Turkish males, serving as an "informal meeting place for men where business can be transacted and news and gossip exchanged." Though teahouses are now found all over Turkey, they were originally an urban phenomenon. One scholar writes, "The transition from [private] guest room to [public] coffeehouse parallels the transition from the isolated peasant village of traditional rural society in Turkey to the commercially conscious, larger, more urban village characteristic of 'advanced' parts of the country." For migrants, teahouses can become especially crucial communal places, "an extension of people's living room," offering televisions broadcasting the Turkish soccer matches, Turkish newspapers, and communion with coethnics.[11]

I visited several teahouses in Berlin and noted that the men often had a cigarette and alcoholic beverage in hand. There were gambling machines where some men played round after round. If a soccer game was on, most eyes were at least partly focused on the screen. Yet while drinking and gambling occurred inside these establishments, which usually had the blinds drawn so outsiders could not see in, to me it remains an open question to what extent such teahouses are distinct from working-class neighborhood bars or pubs in Germany, the United States, or Britain. Men find work opportunities, or hear of them, through the teahouses, and they appear to serve as sites for easy sociability. While many of the men who frequent them are said to be unemployed, this seems partially because the unemployment rate in Berlin is high for Turks. There is a general stigma attached to the teahouses, however, by many Germans and some Turks, which appears greater than the stigma tied to German bars and social clubs.

"Instead of going to the teahouse, gambling, and drinking alcohol, playing cards, we come [to the pigeon coop] and we are with nature," Abit informed me. "I prefer this leisure activity instead of going to the teahouse." Asim insisted: "You don't go to teahouses if you do this. I don't smoke; I don't drink. I don't want to criticize those people heavily . . . [but] it's not my cup of tea. We have good air here, no smoke,[12] friends come here, we have a barbecue, and it reminds me of Turkey and of my childhood. It prevents you from doing bad activities."

Mehmet, a tall, dark, quiet, and serious-looking 35-year-old man from Ankara who worked in an *imbiss*, was more critical of the character of those who frequent teahouses: "You cannot bring someone from the teahouse to your family." Suat echoed the others, stating, "Should I do drugs or drink alcohol, gamble, go to the teahouse? I don't do it. I spend time here, sometimes people visit me, we have tea, we talk, and we have a barbecue. It's a much better atmosphere." Hasan Uluagaç, a 40-year-old building superintendent with short dark hair, a mustache, and a large belly who shared his coop with Turan at the time, said, as he poured us a cup of tea, "If you don't do this leisure activity, then probably we would go to the teahouse. And we would waste our time."[13] Turan bragged of the tea they served us, "This is the best tea you've ever had. It's not like the tea in the teahouses."

At first I was surprised that many of the men drank tea ritually (often in the same simple fluted glasses that teahouses use) while they socialized and tended to the birds, yet criticized men who visit teahouses as wasting their time and living immoral lives. This contradiction only grew when I discovered that, even if infrequently, some of the caretakers did visit teahouses, and all of them had close friends and relatives who frequented them. Mehmet, the most critical of the teahouses, admitted that he might go if invited. Israfil said it does not bother him if people go to the teahouse, and that he goes sometimes as well. I came to see that the *idea* of the teahouse was more offensive to the caretakers than the physical teahouse. The important distinction is that the teahouse was not the place in which the caretakers grounded their identity. Suat said his coop by a canal was a better environment, and Mehmet commented on the "good atmosphere" of his coop in Kreuzberg. Asim claimed his coop next to an auto garage had "good air" and reminded him of Turkey. Timur and

Photo by Oliver Hartung.

Abit felt grounded in nature with their pigeons, even if surrounded by glass towers or concrete. Ali said that when you are with the pigeons, "it doesn't matter if you are in Turkey or Berlin"; and Hasan said, "It's special because we have such a nice place, a garden. You can hang out here, and you have good people. Sometimes we even sleep here, have a barbeque here. You can invite families to this kind of place." Indeed, Hasan's coop was a homey place. It sat in a courtyard behind lush grass, flowers, and rows of vegetables. It even featured lawn ornaments and a small manmade pond, which his birds lazily bathed in. His wife and children came and went, sitting for a while with us as we drank tea in lawn chairs. In my time in Berlin, Beyhan and I also attended a festive barbeque at Israfil's large, fenced-in plot.

These were the places that were central in the lives of the caretakers. As illustrated in the chapter on Piazza San Marco and Trafalgar Square, places can be crucial to meaning making by reinforcing and even structuring interactions. Linking the places these men created with the narratives pertaining to their innate love for animals, it becomes apparent that the coops were the portals that enabled the experience of a connection to nature. The urban environment seemed to melt away as caring for the birds, socially defined through collective practices, transported the men to a realm of experience outside of Berlin and the workaday world. This antiurban sentiment seemed to be part of what led to a negative portrayal of the teahouses.

The caretakers held a perception of nature, encapsulated in Timur's claim that animals only kill "to survive but humans kill . . . for their own interests," that seems to resonate deeply in urbanized society. Nature, the historian William Cronon notes, appears today as the "unfallen antithesis of an unnatural civilization that has lost its soul. It is a place of freedom in which we can recover the true selves we have lost to the corrupting influences of our artificial lives. Most of all, it is the ultimate landscape of authenticity." Indeed, the "culture of enchantment" that James Gibson says is being popularized around nature grows out of the shared belief that nature is a "source of goodness and meaning" through which we can "escape society's encumbrances" and materialism. And in his ethnography of an English exurban village that he calls Childerley, Michael Bell found that villagers believed nature offered a "moral rock on which to stand" that was "free from the polluting interests of social life."[14] Living close to the land, villagers thought, was a more authentic, traditional, and morally superior way of life.

Regardless of whether or not they consistently avoided the teahouses, the (re)connection to nature that the caretakers experienced in Berlin appeared to be interpreted as a more authentic (i.e., traditional) way of embodying their *ethnicity*. They narrated an idealized relationship to nature through their tumblers that they apparently felt was uncorrupted by social interests, yet they also claimed that connecting to nature and loving animals is a timeless, cultural trait of Turks. Through caring for pigeons in an urban setting, they experienced a connection to the natural realm that in turn grounded them in their homeland. This simultaneous anchoring in nature and nation became a lens through which they divided up their social world.

I interpret the men's criticism of the urban teahouses and those who frequent them as a logical extension of the collective framing of pigeon caretaking as the continuance of a genuine Turkish tradition. They spoke of the human-animal bond in near-sacred terms, while the teahouse was interpreted as the antithesis of the natural—and by extension cultural— world in which the men grounded their Turkish identities. Men who found their communities in teahouses were, in the eyes of the caretakers, immoral at least partly by virtue of not embodying traditional Turkish culture. Grounding their identity in the pigeon coop meant grounding their selves in a "pure," "natural" space. The social ties the caretakers formed, and the ways they collectively defined their lifestyle, reinforced their ethnic identification and connection to Turkish culture.

Pigeons perhaps offered more readily available or defensible moral narratives than other leisure pursuits the caretakers could take up. As minorities and migrants in Germany, this dimension took on particular significance. The men recognized teahouses as a part of male Turkish life, but by identifying with pigeon caretaking, they did boundary work that enabled them to avoid falling into a well-known, negative, and stereotypical identity of Turks in Germany and to simultaneously connect to an older cultural narrative. "Just as there is guilt by association," sociologist Gerald Suttles notes, "so there is stigma by location."[15]

I was reminded, however, that there is a moral ambivalence here. When I first returned to Berlin and spoke to Ahmet Dede, a pudgy, boyish-looking man in his early 30s from Istanbul, he lamented, "This leisure activity is actually a waste of time. You don't earn money doing this. I would be happier, for example, if I studied—if I was in your place and interviewed you instead of you interviewing me. I would be a happier person if I studied instead of taking care of pigeons, working in . . . the *imbiss*; and I don't want my child to pay too much attention to pigeons." The men tended to see their obsession with pigeons nested within their stories of thwarted mobility, and they had higher hopes for their children. This led some of them to try to keep their children away from the coops. Mehmet said he once gave away a chocolate-colored bird because his son liked it so much and Mehmet did not want him to get too attached to the pigeons.

Feelings of personal failure and marginalization were common. Ahmet Çabakçor used the words "alienation" and "disenfranchisement" to describe his experience in Germany. Sighing "I have no more energy for this," Ahmet said he had already bought a house in Turkey and was plotting a homecoming. Turan moved back soon after I left Berlin. Almost all of the men discussed the difficulties of securing a stable job, few if any of them seemed satisfied with their economic and social situation, and over half said they often considered returning to Turkey. Mehmet summed up his experience this way: "I regret coming to Germany. It is like a machine. You have no social life. It is not a good life." The men's animal practices, then, should not be interpreted as helping foster successful integration into German society but rather as offering comforting rituals that assuaged—and perhaps momentarily transcended—a harsh reality.

Primordial Ties and Cultural Practices

In the last chapter, I contrasted the New York flyers' relationships with their birds—which I said were driven by social impulses and embedded flyers in a distinct social world—with the "transcendent" cross-species encounters that environmental scholars often highlight—which they say are driven by a personal need to connect with the natural world and to escape society. While these appear to be two distinct—even antithetical—modes of experience, the Turkish caretakers show that they are not mutually exclusive and can in fact be complementary. On one hand, the caretakers felt a sense of kinship with pigeons, revered nature, and said that their pigeons grounded them in the natural world. On the other hand, the kinship they felt was through the social categories of ethnicity and nation, they framed their love of nature as a "Turkish" trait, and they said that their pigeons grounded them in their homeland. Like the New York flyers, their pigeons structured their sense of self and relations to others and anchored them in their neighborhoods. And, similar to the way that the old-timers in New York clung to their coops as nostalgic relics of the communities they lost to the passage of time, the Turkish caretakers saw their coops as sentimental vestiges of the communities they left behind through emigration.

Nature writers have framed kinship ties with other species and the veneration of nature as a kind of new-age totemism, harking back to premodern indigenous cosmologies in which clans adopted plants or animals as their symbol and considered them sacred. These scholars interpret totemism as nature worship, grounded in a belief that "animals and people [were] literally family." By reconstituting sacred ties to nonhumans, James Gibson contends, people construct an image of nature that "is allowed to exist on its own terms, for its own sake, valuable simply because it is there."[16] While such a conception of enchanted and autonomous nature does seem to capture the way that a number of people—particularly environmentalists—interpret nature today, it downplays the social foundation of totemism and seems to deny the possibility that the desire for communion with nature can be rooted in a desire for communion with society.

Many environmental researchers celebrate anthropologist Claude Lévi-Strauss's famous dictum that animals are "good to think with" because it signifies the close relations that indigenous societies had with nature. However, Lévi-Strauss refuted the notion that "primitive" people's relations to nature were somehow less contaminated by social interests than our own and that adherents of totemism conceived of animals as literal kin. Other species, he argued, were central to totemic myths largely because they were useful symbols for expressing the relationship between self and society. Animal kinship was figurative, a handy nomenclature for organizing social relations. Similarly, sociologist Emile Durkheim's analysis of totemism among Australian aborigines led him to reject the idea that "primitive" people actually worshipped nature or valued it as an autonomous entity. Each totem species was sacred, rather, because it embodied an "objectified and mentally imagined" image of the clan; worshipping it was a way to express social solidarity and sanctify the collective.[17] This framing of totemism, though in a diluted and somewhat more secular form, captures an important dimension of the Turkish caretakers' experience of their pigeons. The men's relationships with their birds organized their conceptions of themselves and their social relations. And in venerating tumblers and nature, the caretakers consecrated nation. This interpretation does not dismiss the men's moral elevation of nature and their belief that their attraction to tumblers is "in the blood." Rather, it shows how the desire to commune with nature can be catalyzed by the social self and how the sanctity of close relations with nature can sometimes result from their unique capacity to augment social life.

Like the caretakers' bonds with each other, they felt a "natural" connection with tumblers based on the belief that the birds were of the same ethnic and national category as themselves. The men felt an ineffable, tribal affiliation with the birds because the blood of both species was believed to originate in a common soil. They formed an attachment toward the entire stock of birds, more than any one bird, because it was the *bloodline*—as a Turkish breed—that counted most for the men.[18] It is in this regard that we can conceive of their attachment to tumblers as *primordial*, in the sociological sense employed by Edward Shils and Clifford Geertz. Primordial ties "stem from ... the assumed 'givens' of social existence," Geertz wrote. "One is bound to one's kinsman ... as the result not merely of personal affection, practical necessity, common interest, or incurred obligation, but at least in great part by virtue of some unaccountable absolute import attributed to the very tie itself." The "congruities of blood, speech, [and] custom" can create a sense of solidarity and shared destiny among people linked through kin, ethnicity, and nation even if they do not personally know one another.[19]

Primordial ties are the kinds of attachment that sociologists have long said typify "traditional" communities, but that many argue were disrupted by the social and economic forces of modernization, which gave rise to relations "driven by calculation, interest, and impersonality."[20] As Michael Bell points out, early-20th-century "Chicago school" sociologists mapped primordial ties onto rural communities and depicted the "urban way of life" as one marked by superficial, transitory, and heterogeneous relations. While subsequent urban studies recast the city as a mosaic of "urban villages" within which residents commonly reconstituted the primordial ties of the countryside or (in the case of immigrants) their homeland, many people nonetheless *experience* or *imagine* small-town or pastoral life as more traditional and communal than city living.[21] The caretakers' sense of primordial attachment to tumblers was steeped in their nostalgia and longing for traditional community in both senses: caring for pigeons tethered them to their romanticized childhoods in Turkey, which they wistfully recalled as characterized by durable and affective ties between kinsmen as well as meaningful connections to nature (even in "urban villages"). By keeping a historically "Turkish" pigeon breed, establishing coops in their new surroundings, and forming a peer group with other Turks who also kept pigeons and disdained Berlin's teahouses, the caretakers reconstructed idealized "traditional" communal aspects of their homeland. Critics can rightly point out that the men's primordial frame is itself "socially constructed," that there is nothing that *essentially* makes tumblers and caretaking "Turkish." But the caretakers *experienced* the bond—and ethnicity itself—as essential, eternal, and "natural." And this primordial sentiment oriented their action.[22]

Though the caretakers' animal practices were saturated with their understandings of Turkish ethnicity and culture, there were many facets of their everyday life that they did not interpret through these lenses. Only four of the men, for instance, viewed their difficulty in securing or maintaining a foothold in the German economy as the result of discrimination. And it is notable that, while over half of them had other pets (most often dogs), they did not narrate those relationships through an ethnic lens. Earlier quotes indicate that the men believed Turks love animals in general, yet they did not refer to these other animals' bloodlines or discuss their specific histories and territories of origin the way they did with tumblers. Rather, their pets were interpreted and enjoyed as companion animals with individual biographies and names.

In Turkey, being "Turkish" is an "unmarked" ethnic or national category; likewise, keeping pigeons is (according to the caretakers) a taken-for-granted animal practice. Upon migrating to Germany, the men, like all minorities, became the "marked" ethnic/national category. In such a context, the men's behaviors that differed from the cultural habits of Germans also became—for both them and others—practices that marked them as Turkish. Having a companion dog in the city is not a marked, or foreign, practice in Germany; keeping tumblers is. This was brought home in the common occurrence of curious Germans, including dog walkers, who happened upon the coops and marveled at the frantic tumbles of the pigeons. Such chance encounters usually resulted in amicable interactions between the Turkish men and ethnic Germans, but in each instance the Germans asked why the men kept pigeons. Every answer the men gave highlighted the origins of the birds or the animal practice. Such discussions were never had about people's pet dogs, as keeping dogs is a taken-for-granted animal practice.[23] Though Beyhan and I asked the men to describe the significance of caring for pigeons in general terms, eschewing leading questions about ethnicity, it is logical in this light that understandings of ethnicity would be salient in accounting for pigeon keeping. Yet such accounting, in turn, played a role in cementing a sense of primordial attachment to the birds.

The tumblers, however, were not simply cultural symbols. Like the New York flyers, the Turkish caretakers gained satisfaction in their mundane interactions with the birds, through raising them and through the simple aesthetic appreciation of watching them in flight. Some kissed the birds, and Turan tenderly spoke to a sick baby pigeon as he fed it special food through a funnel. Suat told us, "You always wonder whether this day the tumbler will be in good shape, how many times it will tumble, how far it will fly . . . [and] we are curious with a baby, about what color it will be. They are very enjoyable for this." Some of the men referred to caring for pigeons as an "escape." Abit noted, "It helps us recover from stress, from the pressures of everyday life. We come here one or two hours after work—this is very healthy for us." Ali informed us, "I go to work, and after work . . . I go to the roof and spend the rest of my time with the pigeons. You relax. You forget everything. Your mind approaches zero . . . I am very calm, and I learned it from the pigeons." Sinan also said that pigeons serve this function: "Sometimes you need an escape, and this is my escape. If I interact with tumblers, I forget boxing. I also walk my dog and spend time with the chickens, but tumblers are the best for forgetting boxing." The escape that the tumblers afforded was not a flight from

Mehmet finds sanctuary in the coop. Photo by Oliver Hartung.

society altogether, but rather a temporary respite from tedious routines and the estrangement of living in a foreign city.

It is crucial to situate the men's relationships with pigeons within their life trajectories, social relations, and context. Each man told a story of migration and disruption, a story wherein a rediscovery of tumblers forged a tangible link with home. It was not a casual endeavor, as properly caring for pigeons required several hours per day and a few hundred dollars per month. The men told stories of bouncing around from city to city, chasing jobs and being unemployed, but then finally settling down in one city, starting a family, and gaining some level of economic well-being (at least compared to their unemployed peers). It was largely at this moment when the men were able to involve themselves in a pursuit like caretaking that requires social stability and a bit of disposable income.[24] But they narrated this moment of rediscovery as a reawakening.

In the caretakers' youth, keeping pigeons in Turkey had a taken-for-granted quality. Years later in Germany, the men described an intense, visceral reaction to seeing tumblers again, which threw them back achingly to the comfort of the familiar. The communion with these animals recalled and partially recreated the earlier context in which they cared for pigeons in their homeland. Handling the birds revitalized the connection in a conscious way. Orhan's story was typical. He told of how just a few years ago he did not have tumblers and did not know that men in Berlin kept them, but he saw some in the air while walking down the street one day. He dramatically recounted that he felt like his heart was going to leap out of his throat as he longingly remembered his childhood spent with the birds. Now the birds were a central part of his life, and he helped other men secure spaces and their own birds.

The caretakers were either first-generation migrants or the children of immigrants. This means, especially given Turks' history in Germany, that the men could not escape their ethnicity in the eyes of others even if they wanted to. Yet they still exercised a limited sort of "ethnic option," to use sociologist Mary Waters's phrase,[25] by consciously and symbolically staking their identity in the nonessential practice of caretaking. It would be inaccurate to assume that the men simply reproduced their ethnicity through pigeons in Berlin. As these men from diverse regions in Turkey came together as a miniature diaspora in Berlin, through their animal practices they jointly constructed a unique definition of "Turkishness" that reflected both their individual pasts and their present collective reality. In doing so, they renewed their Turkish identity in a self-directed way. They sewed up a fissure created in their ethnic heritage through migration, and they created sheltered, sociable spaces where those who shared their passion could congregate.

Despite the absence of contemporary sociological scholarship on the subject, it is not only "primitive" human groups who forge collective identities and social ties around other species. Further, the caretakers indicate that studying immigrants' animal practices could provide unique insights into the processes of ethnic and cultural reproduction and adaptation.[26] Lastly, the caretakers integrated two modes of experience that are too often regarded as contradictory: their pigeons were both a portal to a more natural place and an anchor for the social self. Though nature writers often frame the felt primordiality of kinship with other species as premised on the imbuement of nature with an otherworldly significance, the caretakers suggest that, at least sometimes, the primordiality of cross-species ties can result from locating nature within hallowed social categories like ethnicity and nation rather than viewing it as autonomous from the social realm.

FIVE

Joey's Brooklyn Pet Shop

Cosmopolitan Ties in a Changing Urban Landscape

BROADWAY PIGEONS and Pet Supplies, known to regulars as "Joey's" or "the pet shop," was nestled in a blighted part of Brooklyn, bordering on largely Latino Bushwick and mostly black Bedford-Stuyvesant.[1] Though only a handful of pigeon stores were holding on in the five boroughs as rooftop coops vanished, Joey's shop was new. Opened in 2002, its location reflected the hobby's changing ecology. It seemed that Bushwick was the only neighborhood left in New York that had a lot of "action" in the sky. The faded rusty trestles of the elevated J and Z train lines hulking over Broadway cast a shadow on cheap clothing stores, abandoned buildings, trash-strewn lots, check-cashing centers, delis, liquor stores, and storefront churches. On Sunday, a parade of African Americans in their church attire added a touch of class and color to the drab scene. A different weekly custom took place inside of Joey's storefront on this day of rest, where a racially mixed group of working-class men—many decidedly not in their Sunday best—gathered to hang out, swap stories, and offer evidence of who kept the best pigeons.

The confluence of white, black, and Hispanic flyers of varying ages was an organic result of inner-city demographic shifts. Though discord accompanied racial succession in Brooklyn's former working-class white neighborhoods such as Bushwick and East New York, which were historically characterized by provincial and ethnic ties, I discovered that some of the ethnic white flyers, who remained after their cohort's suburban exodus, passed on this animal practice to some of the Hispanic and black males who moved in decades ago.[2] Although most flyers I knew who lived through these neighborhood changes still harbored primordial loyalties and commemorated ethnic enclaves, their impassioned identification with pigeons compelled them to traverse racial and generational boundaries that were still salient in other social spheres of their lives.

While the Turkish caretakers drew from a shared ethnic and cultural background to frame their animal practices and foster group solidarity, the New York flyers' interpretations of pigeon keeping were locally improvised, and achieving a "we-feeling" involved a delicate balancing act of bracketing differences considered irrelevant to group practices while creating social distinctions within the group based largely on perceived "natural" disparities in the quality of their pigeons. As the central node of flyers' social world, the pet shop was the epicenter of these negotiations. For men like Carmine, the flyers they encountered there served as their *primary group*, defined by the pioneering sociologist Charles H. Cooley as a collective "characterized by intimate face-to-face association" that is "fundamental in forming the social nature and ideals of the individual."[3] Through shop talk at Joey's, flyers were socialized into the group's norms, tastes, and stock of knowledge. As discussed in chapter 3, even when alone on their rooftops the flyers obsessively compared themselves and their birds with each other. In light of this, the pet shop was the arena in which the flyers vied for social status face-to-face.

Because each flyer invested his esteem in pigeons, in the pet shop he was often under a "psychological burden to perform for his peers—and for himself."[4] In presenting, affirming, and challenging one's status through "ballbusting," the men oriented their social self toward brash displays of masculinity. Though many of the men did in fact evince some degree of emotional attachment to their birds on the rooftop, pet shop interactions endorsed an ethos of not "giving a fuck" about one's pigeons. The pet shop also offered opportunities for the flyers to validate the repetitive and dirty work they put in each week to maintain the coops and care for the birds. Yet the men got together not only to compete. They cared about and kept up with each other. Their affinity, forged through pigeons, diffused beyond their animal practices.

While the last chapter showed how pigeon keeping reinforced primordial ties and an ethnocentric perspective among the Turkish caretakers, this chapter illustrates how pigeon flying in New York fostered what I call *cosmopolitan ties*—that is, "informal, voluntary, and affective" local attachments that transcend the master statuses (e.g., race, kin) that traditionally serve as the basis of place-based primary groups.[5] Within this accidental community formed in response to changing neighborhood ecologies, the pigeon came to function as a sort of allegorical totem that, by standing *outside* of any particular ethnic category, united anyone with an interest in and commitment to pigeons on neutral turf. Particularly as gentrification increasingly threatened this already moribund hobby, it seemed that the flyers needed each other more than ever to sustain the practice. More than that, though, I found that it was largely through immersing themselves in a social world based on pigeon flying that the men found an opportunity to "be somebody."[6]

The Social Life of the Pet Shop

Over the doorway of an otherwise nondescript storefront bathed in fluorescent light, a large green awning broadcasted "Broadway Pigeons and Pet Supplies" and advertised some pets and products that it seldom carried, such as fish and parrots. While Joey carried the latest pigeon products and replenished his supply of pigeons through weekly auctions in Long Island, he showed little concern for meeting expectations of a full-service pet shop. The entire left side of the store was given over to hundreds of neatly stacked 50-pound sacks of pigeon feed, which sold for $15–25. The counter on the right was filled with medicines and health supplements for pigeons. Next to the counter was a cash register, though Joey preferred to do business out of his pocket, and a surveillance monitor. Leashes, litter, and some cat and dog food filled in the remaining shelves. In the winter, Joey sold hay so flyers could prepare their pigeons' nests for breeding. Also for sale were the long bamboo rods that the men used to chase their pigeons. The only decorations were two identical posters that displayed drawings of fancy pigeons, and several pictures: Joey's infant son and stepdaughter; a smiling young man attached to a news clipping showing pigeons released at his funeral; and a man next to his coop between the scrawled phrases "Mikey's Outlaws" and "Sky is mine."

A large sign on the back wall announced in bold print "IN GOD WE TRU$T. ALL OTHERS PAY CASH! DO NOT ASK FOR CREDIT!" Despite this stern warning, Joey kept a tab for about two-dozen regulars. A Plexiglas window afforded a view of the back room, where birds for sale were displayed. It had the musty smell of pigeon dung, and dander blew around thanks to an encrusted fan. On the left were four simple pens filled with dozens of pigeons that sold for $5 each. On the right were stacks of smaller metal cages that housed higher-valued pigeons ($10–25 each). Joey kept a grimy container of Rite Aid chest rub on top of these cages, which he used to polish the fancy pigeons' beaks and feet to enhance their marketability.

Photo by Dave Cook/Eating in Translation.

Though Joey's store was not well suited as a gathering place, about 40 to 50 regulars cycled through and hung out on Sundays between the store's hours of 9:00 a.m. and 1:00 p.m. At any given time, as many as 30 men ranging in ages from the teens to the 80s would be sitting on bags of feed, standing outside or by the register, and inspecting birds in the back. Joey often bought doughnuts for whoever dropped by, and sometimes regulars bought rounds of coffee.

COMPETITIVE SOCIABILITY

Sociologist Elijah Anderson notes that an individual's "personal sense of rank and identity is precarious and action-oriented." Given this, the pet shop was a place where reputations were "cast, debated, negotiated, and then defended."[7] Pet shop interactions were marked by what I call competitive sociability, as flyers jostled for social position and esteem while hanging out. Caring for pigeons took a lot of work, and they staked their self-worth on the quality of their birds and coops. This lent the constant "bullshitting" in the pet shop a jagged edge. As men, they had to be able to withstand having their "balls broken," yet they could not afford to let accusations and insults linger as representing even a partial truth. This was brought home in an incident at the pet shop involving Carmine and his much younger white friend named Fat Charlie, who visited him on his roof and sometimes gave him a ride to the store. One fall afternoon, Fat Charlie was ribbing Carmine in front of other flyers about a time that he was on Carmine's roof and Carmine would not chase his pigeons. Carmine defended himself by saying that he had arthritis. But Fat Charlie egged him on, calling Carmine's reason an excuse and insinuating that he was afraid to lose his pigeons by chasing them. Carmine became infuriated.

As one of very few people his age who was physically able to keep pigeons, Carmine was usually immune from such attacks. His flying reputation threatened, Carmine was determined to prove to this person less than half his age that he was the bigger man. Carmine called Charlie all sorts of names like a "fat fuck," and he later recalled, 'He was about ten feet away, and I was already sizing him up . . . I was gonna punch his fuckin' lights out. I swear, I was gonna punch him right in the teeth and he would have went down, boy.' Carmine got right in Fat Charlie's face, who seemed as surprised as many of the regulars at the intensity of Carmine's rage, and threatened him. As the men pushed the two apart, Carmine stormed out of the store and gave his farewell: 'Go fuck yourself!' Five years later, there was still no sign of rapprochement.

Erving Goffman argued that when "disruptive events" interfere with one's presentation of self, such as Fat Charlie's continued attack on Carmine in front of an audience of his peers, "the individual whose presentation has been discredited may feel ashamed . . . experiencing the kind of anomy that is generated when the minute social system of face-to-face interaction breaks down." And sociologist Jack Katz notes that "anger emerges from a falling out of community."[8] Ballbusting could produce anger not only because of its power as metaphorical emasculation, but because it threatened to strip a man of his social standing in the group.

While ballbusting could turn serious if a man's reputation was impugned, tense interactions could quickly be defused when a man was affirmed. On one occasion, a middle-aged Puerto Rican named Fernando publicly doubted Carmine's ability to train flights to route. Carmine said tersely, 'Get the fuck outta here! You don't know a fucking thing! When you've been flying birds as long as I have, then you can talk to me about routing!' Fernando walked away, and as he turned around the audience braced for an escalation in tension. But Fernando cracked a smile and said, 'Yeah, you been flying a long time! I can't touch that.' Carmine's scowl melted, and they chuckled as they hugged each other. Fernando constructed a "narrative ending" that "recast the meaning of the entire interaction"[9] as a validation of Carmine's expertise and experience. This changed definition of the situation transformed tense voices and muscles into intimate hugging and laughter, dissolving competition in favor of sociability.

Pet shop interactions fluctuated situationally, and often artfully, between the poles of competitiveness and sociability. As the men probed the limits and contours of ballbusting, they learned how such interactions could be used simultaneously to bring the men together for the pleasure of association and to shore up one's reputation. Usually, these interactions were less tense, and more playful, than the two above incidents Carmine was involved in. They often consisted of what Douglas Harper calls "abusive joviality," which is a common ingredient in interaction among well-acquainted men—particularly working-class men.[10]

A middle-aged Puerto Rican named Mikey was once featured in a glossy "hipster" magazine called *Swindle*. While the men passed around and admired the story one day in the pet shop, a younger Puerto Rican bodybuilder named Swappy yelled, 'Gimme that book!' Opening it to a centerfold picture of Mikey's coop, Swappy joked, 'Look at that shit! I would be embarrassed! If I was you, I would buy up every copy of this book and burn it! You will never see garbage like that on my roof!' As we all laughed, Mikey had a stinging rebuttal that pointed to their intimate relationship: 'Half of those birds are yours! I got those birds from you!' As the audience of 15 laughed louder, Swappy returned fire. 'Guys, I'll tell you. If you came up on my roof and found one bird like this, I'll give you my whole stock!' Pointing to a close-up photo, he chortled, 'What is this, a crow? Damn, those shits is ugly!' Mikey, who according to the article had been flying for 40 years, retorted, 'My birds fly, man! And if you look at a map of pigeon flyers in Brooklyn, you ain't even on the map!' The crowd, including a cadre of old-timers who particularly valued experience, responded, "Oohh! Ouch!"

Mikey seemed to have succeeded in defending his birds and his status. Manny chimed in, challenging Swappy, 'Let's see some fuckin' pictures of your birds. Where's your magazine? Leave the man alone. His birds fly.' Swappy responded, 'I'll tell you the truth, they do fly. But when they . . . taste that thin air in the stratosphere, they go "phrew!" [He made a scattering motion.] They can't handle it!' Swappy pulled me in as a witness: 'You been on my roof. I got nice birds?' I silently nodded yes. He went on, 'My birds go. They get up there in the [clouds] and you can't even see them, it's like a pencil dot. I chucked up with Macho, LG, and some other guys—I gave 'em a half hour head start and I still beat them niggas! We chucked from downtown Brooklyn by the clock.' Swappy's words were given credibility when LG conceded, 'He beat me. He did.' Swappy continued, laughing as he pretended to make a wager, 'I'll put $100 right now! I'll chuck my birds. I'll take 'em right off their eggs!'

Swappy and Mikey then mock-wrestled each other as the pet shop swelled with mirth. The insults parried indicated the depth of the men's friendship and played on themes understood by every flyer. In front of an audience, Mikey and Swappy affirmed their own status as flyers. Everyone present was impressed with Mikey's magazine spread. Yet Swappy also came out ahead, for LG and I validated his claims that his birds fly well.[11] Thus, competition among the flyers was not necessarily a zero-sum game. Through sociability, the men could work through social relationships and status disputes in a playful fashion.[12]

EXPERTISE AND MANLINESS

In the pet shop, it often seemed that being a good flyer meant being a "hard-ass." And no one seemed to embody the hard-ass more than Michael Scott, Joey's bald-headed, muscular big brother, who cursed like the proverbial sailor and reveled in trash-talking. On a particularly frigid February morning, Mike came into the pet shop as most men were complaining about the temperature. In contradistinction to these "pussies," Mike bellowed, 'Anybody want to see some fucking birds fly?' There did not seem to be any takers, but Mike went on. 'Come up on my fucking roof right now if you want to see some birds rock!' Silence filled the room, as no one seemed enthused about being on a rooftop when the wind chill was around zero degrees. Then Mike added an absurd ultimatum: 'They'll fly for an hour and a half solid. If they don't I'll suck my own dick!' Mike looked everyone in the eye until they looked away, but when he got to me I withstood his gaze. He barked, 'Let's fuckin' go, bro!' So, in sticking with my research principle of never turning down a participant's invitation, I found myself shivering on the roof of his Bushwick Avenue property. The water canisters were frozen solid, but he pushed his birds into the clear blue air and whistled as they ascended.

I had to give it to Mike—the birds were "rocking." They flew tightly together, doing quick turns and dives in unison and alternating between the formation of a ball and a ribbon. Mike replaced the frozen water while

I waved the bamboo pole vigorously, as much to keep my hands from going numb as to chase the birds. They looked magnificent bathed in the winter sun, swirling in front of the distant Manhattan skyline with no apparent leader, like a school of fish. But it seemed that Mike's "partner" Tito or his brother forgot to feed the birds that morning. Mike said there was no way the pigeons could fly for long if they were starving. Delighted to get off the roof, I did not remind Mike about his promise to perform fellatio on himself.

The length of time that Mike's birds flew was less important than the fact that he took action in front of his peers to back up his presentation of self. Because status "is a pattern of appropriate conduct," not "a material thing to be possessed,"[13] there was a need to occasionally put one's money where one's mouth is in the pet shop. Such moments became group lore that added credence to the front that flyers performed in the store. In one memorable event, a young Hispanic man named Isaac parked in front of the store on a cold January afternoon with several crates of flights. Rather than bring the pigeons into the pet shop to sell, he walked farther away until he was no longer under the elevated train. Slowly, the flyers realized what was going on: 'Oh shit! He's gonna toss the birds!' We all ran outside just in time to see Isaac "chuck up" the pigeons, releasing them from the crates. As the confused pigeons circled over Bushwick Avenue in two separate bundles, three or four excited men ran to their cars and sped off to their coops, hoping to catch strays as the birds tried to return to Isaac's coop. Isaac, maintaining a blasé demeanor, entered the store even as the rest of us stood outside and watched to see if the pigeons would orient themselves and begin going home. The other men's remarks about how difficult a time his pigeons would have running the gauntlet through Bushwick only seemed to bolster his esteem. Isaac defiantly retorted, 'I don't give a fuck! If they don't make it back, I don't want them.' This event was talked about for months afterward.

Flyers were generally rewarded with affirmation when they indicated that they "don't give a fuck" about losing birds. The implied opposite was that one was afraid to lose birds, that one was attached to his pigeons like pets. Though flyers' actions on the roof indicated that many of them *did* form caring and emotional attachments to their birds, in the pet shop they coded such ties as feminine. The "I don't give a fuck" frame also fostered the illusion that the men *chose* to lose birds when, on the roof, they actually lamented these losses and sought to avoid them.[14]

It was common for the men to call someone out for indicating attachment to pigeons. When a black man in his late 30s named Manny saw a few flyers buying hay in the store one winter morning, meaning they had locked up their birds for the season to breed, he accused them of acting out of fear that they would lose their birds to predators (hawks went after flyers' birds more in the winter because other prey hibernated). With his trademark jocularity, Manny invoked the tone and cadence of a preacher to impugn their motives: 'There comes a time when you got to let them birds out! *There comes a time!* You gotta open up them cages! *Fly the birds! They got wings*! Don't be afraid of the hawk!' The men who continued to fly through the winter laughed cathartically, distinguishing themselves—for the moment, at least—as the model for "real" flyers. Those who started breeding "early" did not vocally defend their actions, validating the morality attached to Manny's sermon. When a muscular and dapper Puerto Rican regular uncharacteristically voiced regret over losing a 'nice red magpie' pigeon, LG chastised him, 'Who gives a fuck?! If it ain't on the roof, I don't want it anyway.' Teddy the Greek, a white old-timer, supported LG's stance: 'A flight is only yours when it's in the coop. I don't give a fuck. That's what this is all about.' Terry, a 35-year-old black sanitation worker, had this to say about his neighbor: 'He just likes to fly his birds over his roof and watch them. He won't roll out. He a *dog catcher*. And that's my friend. He likes to toss the food from the can and the birds land on his arm.' As some of the guys around Terry chuckled, he crowed, '*My birds are afraid of me.*'

Though the men varied in the degree to which they acted like they "don't give a fuck," some bought into the public performance of this attitude to what seemed like an extreme level: valorizing violence toward their birds. Violence was to some extent unavoidable in "the pigeon game." Because there is a range of communicable diseases that pigeons can easily pass to one another, sick pigeons sometimes had to be "culled" (killed). Most of the men had a hard time with this, and they inoculated their birds to try to prevent disease. Carmine was ashamed to describe how he drowned sick pigeons, and said that he had to turn his head away. Even Manny, often prone to manly boasting, drew the line at killing. With a smile, but using a pleading tone, Manny asked his friend, 'How you gonna kill the bird, man?' He added, 'If mine get sick, I just drive 'em for a while and release 'em.' But Manny's friend also seemed unwilling to have blood on his hands, explaining that he fed sick pigeons to his dog. Manny protested, laughing, 'But you an accessory, man!' LG, a burly Puerto Rican in his 30s, shook his head at these "soft" men. Cocking his head sideways to imitate a sick pigeon, LG boasted, 'As soon as I see them like this, I kill them motherfuckers! I don't give a fuck!' Big Frank chimed in, 'I killed a hundred the other day! Drowned em.' Although those who publicly acted as if they derived pleasure from these mercy killings were in the minority, I did not see other flyers challenge this cavalier stance. In fact, like Manny, even those who did not have the nerve to cull their pigeons often found humor in others' ostensible callousness. Publicly demonstrating consternation would betray gender-inappropriate attachment toward the birds.

A person like Big Frank, however, also demonstrated the limits of leaning too heavily on the "I don't give a fuck" trope to build a reputation. Decoupled from a demonstrated diligence in breeding and training one's birds, this attitude could evoke disapproval.[15] Big Frank was an Italian American in his mid-60s. Yet he dressed and behaved more like some of the young men of color in Bushwick, where his coop was, than like the Italian old-timers of his cohort. He wore flashy jewelry, which he sometimes offered to sell, and a hooded sweatshirt emblazoned with the face of the deceased Bed-Stuy rapper Biggie Smalls. He claimed he would not date "white chicks," and he once told me, 'If she's over 28, my dick don't work.' Big Frank actually lived in Long Island, but he drove his black Cadillac Escalade with 22-inch rims, or his red Mercedes convertible, to his coop on top of a tenement in Bushwick. His SUV had a large sticker on the back that read "Cash 4 Gold," indicating the business that he was in.

Big Frank epitomized the "I don't give a fuck" attitude. He would fly the whole stock at one time, which was too unwieldy to control. He would sometimes even buy another flyer's entire stock and roll it out as soon as he got it home, leaving the birds no time to adapt to new surroundings. And when he lost a lot of birds, he simply bought more. In terms of sheer numbers, the results were impressive. Big Frank usually had over 1,500 flights. From my apartment, I could often see his massive bundle dwarfing all of the other stocks around and casting an ominous roving black shadow on the pavement below.

Big Frank certainly commanded admiration from the group, as he had been flying for a long time and was renowned for a long-defunct coop he used to have—called High Loft—on a factory roof that reputedly had a television and heating and served as a clubhouse. The men also enjoyed his enthusiasm and endless trash-talking. Big Frank was good company.

Yet some flyers criticized Big Frank's version of the "I don't give a fuck" attitude because he seemed to lack the discipline that the men liked to bring to their craft. Richie, while laughing in delight at how easy it was to catch strays off Big Frank, objected that he "doesn't play right." "He buys a new bird and it's out the next day. He don't care." Richie acknowledged that losing pigeons "is part of the game," but he admonished Big Frank and others like him who "blow their money" on pigeons in this way. Carmine disgustedly recounted when Big Frank lost over 100 pigeons because he flew them after sunset, and said he was "ruining the pigeon game" by paying top dollar for birds. On one occasion when Big Frank was in the store negotiating to buy a stock from another flyer, a regular whispered, 'How many roofs has he been on? You can't just buy other people's birds, you gotta raise your own.'

Most flyers treated their pigeons as a long-term project or a job: they carefully maintained the coops; they selected, mated up, and bred birds that performed well; and they subjected the pigeons to training schedules that could stretch for months. Big Frank's reckless manner, on the other hand, echoed his atypical work habits. He was a self-described hustler and partier who celebrated a fast and loose lifestyle. Big Frank seemed to lack the patience to reliably breed or condition his own pigeons. Most men bought a "foreign" bird only if it was an exceptional specimen, preferring their stock to be the product of their own hands, and they knew that chucking a new bird into the air right away was folly. But when Big Frank drove his Escalade into Bushwick from over an hour away, it was not to "scrape shit." He footed the bills for the building and delegated much of the day-to-day maintenance to his "partners." Big Frank came to roll out. He reminded me of an athlete who loathed practice but lived for game day.

For most flyers, coop maintenance and training were part and parcel of keeping pigeons and were a source of pride. While the flyers agreed that one should not cry over lost pigeons, they tried to minimize losses. Flyers secured the most bragging rights, in fact, when they only lost a handful of birds after rolling out and crashing another stock. It was when faced with a desire to prove how dedicated one was to rolling out, which seemed to always be Big Frank's prerogative, that a flyer might construe losing large numbers of birds as positive. But by not bringing the same attentiveness and industriousness to his craft as most of the men, Big Frank's macho script sometimes seemed a bit off the mark.

DRAWING DISTINCTIONS

The pet shop socialized men into the stock of knowledge of the group. They learned how to spot and treat disease, what equipment to buy, how to train the birds, and when to breed. But it was also through the pet shop that the men had their values and tastes shaped. There are standards for every breed of pigeon. While such standards are described in books and magazines, the flyers also had their own local definitions that, while informed by the ideal, varied from it. Just as the men's reputations could be situational, birds became "good" or "garbage" in context. I saw show-quality flights sell for over $100 at auctions. However, most flyers gasped at paying such prices, and they claimed that such pigeons were likely not good flyers. Their idea of a good pigeon combined aesthetic and pragmatic considerations.

The pet shop overflowed with episodes in which the flyers tried to work out the worth of a bird. One time there was a beautiful blue bearded flight in Joey's store. It had already sold for $20, but the buyer had not picked it up. Every visitor to the store admired the bird at length, stroking its head, staring into its eyes, and pulling out the wings and the tail. The men would then ask Joey about it, but most acted disgusted at the high price that it sold for and castigated the buyer in absentia. Carmine said no pigeon was worth that price because it would be flown, not kept for a show, and might be captured by a hawk. Fernando joined Carmine in claiming that he had several birds on his roof that were nicer than that. Perhaps most problematic, several men pointed out that the bird had a white mark above the beak. According to the standard, this was a mismark. Carmine complained to me that 'if you took it to a show, they wouldn't even judge it. It's foul.' However, other men did not seem to care, or even said they liked the mark. Such a mismark was so common that it had a name, "shuster." A few men relied on this name to validate the bird's worth, saying that they liked shusters. Shusters had not yet become a valued variant among most flyers, like beards or caps, but it appeared they were on their way.

Even experienced flyers rarely bought a bird without showing it to others. If they gained consistent affirmation that it was a nice bird, they were more likely to buy the pigeon. Such instances enabled flyers to display their knowledge and flaunt their reputation, as many men expressly sought the advice of old-timers, but such instances also played a vital role in producing the stock of knowledge that the men drew on. Notions of a good bird were durable but not static, as fads—local and national—came and went.

As the men's rooftop activities were made meaningful largely through their social relations, they took local definitions of garbage birds quite seriously and sought to keep pigeons that would impress others. Keeping good birds was also a moral matter, for the men saw these collective definitions, fashioned through a combination of distant history and local folklore, as prescriptions for the kind of pigeons that a "real flyer" ought to have. Owning garbage birds might also signal a lack of expertise and hard work, traits highly valued by the men.

Through definitional negotiations at the pet shop, the social world of the flyers fashioned its own aesthetic sense. Indeed, Darwin wrote extensively on the ways that pigeon fanciers, disproportionately drawn from the lower classes, created new breeds through selecting on their own idiosyncratic aesthetic tastes. These working-class men, most of whom seemed so nonchalant about their own appearance, eschewed markers of high status, cared little for art or museums, and preferred the *New York Post* over the *New York Times*, developed what outsiders might consider a snobbish attitude about pigeons. Like judges at the Westminster dog show, the men inspected each pigeon on the roof and in the pet shop, using a mental checklist. Certain details instantly disqualified a pigeon as a worthy specimen, such as the tiniest hint of black (a "splash") on what was supposed to be a purely orange beak.

Most outsiders could not even discern a feral pigeon from a purebred $100 flight. To have this literacy, a person must undergo "perceptual socialization."[16] For the initiated, though, "street rats" and flights were not even interpreted as members of the same family. Form was often appreciated apart from function for these flyers, and the ability to make such distinctions became a marker of status. This aesthetic sense extended to the performance of the birds as well. The men learned from each other what it was they should expect and desire from their birds in flight. There need be nothing inherently better, for example, about pigeons circling, diving, and routing in unison as opposed to flying in several bundles. But to the flyers the former flight pattern was an unquestioned ideal and a thing of beauty.

Serendipitously simulating totemic logic, which used "the diversity of species as conceptual support for social differentiation,"[17] the flyers used perceived natural differences in pigeons as the basis for drawing in-group social distinctions. One of the fundamental distinctions that consistently divided loyalties in the group was not the flyers' race but rather the "race" of the birds they kept. The New York flight was the traditional rooftop pigeon. About two-thirds of the men strongly identified with this bird and some, like Carmine, chastised those who kept other types of pigeons. But there was a sizable minority of men who kept tiplets, and rowdy but (usually) good-natured debates often erupted between "flight men" and "tiplet men."

Terry, a black man who kept flights, and Swappy, a Puerto Rican tiplet flyer, went at each other in the pet shop one spring day. Swappy took the opening shot: 'If it don't have a black beak [like tiplets] then I don't give a fuck! They should get everybody's flights and line 'em all up and go "Pow! Pow!"' Smiling, Swappy acted as if he was firing a rifle. Terry, the "flight man," retorted, 'I don't like rats on my roof, bro! Those things are street rats!' Terry admitted that tiplets are "smart," 'but I just don't like 'em. I don't like the way they look.' Swappy offered a comical rejoinder that played with the implicit racial dynamics so often lurking in the background at the pet shop, 'Yo, why you gotta discriminate against black beaks, can't we all just get along?' At that, they hugged each other as belly laughs erupted. Terry then bragged, 'Yo, I'm all the way out in Canarsie and my flights go to downtown [Brooklyn].' Swappy indicated that he smelled bullshit: 'Yeah, I seen 'em in the mall when I was in there!'

Whereas flights have orange beaks and white wing tips, tiplets have black beaks and are often gray. It was generally recognized that tiplets could more easily fly in a bundle without getting lost. Those who kept flights framed this as a negative, for it became too easy to manage a stock and too hard for others to have fun by catching strays. They belittled tiplet flyers with the taunt "tiplets are for kids." The implication was that "real men" took up the challenge of training and managing flights. Big Frank, who kept flights, got at this point when he joked, 'They oughtta have a separate entrance to the store for tiplet flyers,' alluding to the traditional ladies' entrance in taverns. "Flight men" added insult to injury by highlighting that tiplets looked more like street pigeons, calling them rats. "Tiplet men," on the other hand, delighted in how "smart" their birds were and often laughed as "flight men" lamented losing dozens of pigeons when they rolled out. They framed their birds as more able pawns in the pigeon wars.

Those who kept flights grounded their legitimacy in a place-based history. As flights were the original, they were viewed, even by most tiplet flyers, as the "true" New York rooftop pigeon. One time a customer in

Joey shows off a juvenile flight that he bred.

the pet shop bragged to Joey about how many quality "fancy" pigeons he had and complained about the quality of the few tiplets and "bronzies" that Joey was selling. Joey simply retorted, 'Bro, this is Brooklyn. Why would I carry that stuff?' Panama, an ardent "flight man," taunted, 'This is *Brooklyn*! Brooooooklyyyyyn! We got flights here! Tiplets are for kids. Joey got the birds he got because those are the birds we fly in Brooklyn.'

Old-timers were more likely to keep flights and to legitimate this choice through appeals to history. They were conscious bearers of tradition. The younger flyers of color were more likely to keep tiplets, but this was likely more of an adaptation to their environment than a repudiation of history. In some parts of Bushwick, it was almost as though a cold war was under way. Men stockpiled more and more birds in order to keep up with the size of other stocks. A few pockets of this neighborhood had 10–12 coops within several blocks of each other, and so aerial tangles with other flyers could be a frequent occurrence. By keeping tiplets, the men were better equipped to handle these battles and lose fewer birds. Most men complained about the cost of pigeons in money ($50–200 per month) and time, and so it was practical for the men to fly tiplets to keep their costs down. As keeping tiplets became more common, a place-based identity began to emerge around these birds as well. Some called tiplets the Bushwick bird. Even in their prime, the old-timers never kept so many birds. They viewed the younger men's large stocks as ostentatious, and were unsympathetic to the practical reasons for keeping tiplets.

A PLACE AT THE PET SHOP

Despite the distinctions that the flyers drew among themselves, simply owning and flying pigeons—especially as the hobby declined—qualified one for a place at the pet shop. The men readily embraced and taught newcomers, donated birds to each other, and made a point of breeding their friends' birds. Tiplets and flights were all "rats with wings," or indistinguishable at best, to outsiders. Flyers also found commonalities in their discussions of work, their home lives, and sports; and they had common enemies, such as "the hawk" and neighbors who filed complaints against them. They shared their ups and downs. When a young flyer named Pasquale Esposito was tragically murdered in 2007, the men simultaneously released their birds in solidarity, and they also released pigeons at his funeral and went to the arraignment of his killer. Ever since Joey had his first child, the guys asked for photos and even gave gifts. Relationships formed in the pet shop led to visits to each other's rooftops, social interaction beyond the rooftop and pet shop, and the occasional "job hookup." To borrow a poignant phrase from Elijah Anderson, the men "care enough about one another to compete."[18] They invested considerable energy into these ties, and so there was an ethic of caring under their rugged talk.

A basic collective understanding of their social world with its own rules, rewards, and morality united the men. Their interactions with animals became a guiding principle in their lives, influencing the social networks they selected into and the ways in which they experienced and interpreted the city. Through shop talk, debates, and reminiscing about classic aerial encounters and the heydays of rooftop coops, the men constructed a narrative in which to situate their activities and their identities. Old-timers like Tony the firefighter or Freddie, a frail cigar-smoking legend, recounted when there were 'five or six coops on one block of Smith Street.' But the younger blacks and Puerto Ricans from the remaining areas with action came in with stories from the front lines of today's "pigeon wars." While the younger men of color complained of losing dozens of pigeons or bragged of capturing just as many, the white old-timers who lived in far-flung neighborhoods complained that they were lucky to catch a single stray. Old-timers lived vicariously through the stories of the younger men, and used these interactions as platforms to recount their own glory days. The younger flyers respected their experience and weathered criticisms from old-timers like Russ, who once called a man who had been flying for 34 years a "newcomer" and barked, 'I've been flying birds since before you were jerking off! And you can't jerk off no more!' Through stories from past and present, whether truthful or "bullshit," the men built bridges that established a continuity between experiences of the mostly white old-timers and the mostly young and middle-aged blacks and Puerto Ricans.

The men shared a fundamental conception of how a "real pigeon flyer" behaves. Thus, they could all partake in the rewards of esteem that accrued to successful performances in the sky and in the pet shop, and they could collectively condemn those who flouted their social and moral codes. Certainly, the men presented a front in the pet shop that did not always match their behavior on the roof. But rather than simply dismissing their impression management as bullshitting, it seems to me that the "mask" through which the men performed in the pet shop represented "the self [they] would like to be" in the eyes of the other flyers.[19] As sociologist Robert Park pointed out, we often publicly perform the self in a way that we privately strive to live up to. In the pet shop, the men endeavored to "exemplify the officially accredited values" of the peer group; their sociable interactions were everyday "definitional ceremonies" in which the flyers "clarified their understanding of their identity by having once more performed it."[20]

It was the flyers' shared activity in caretaking, and each individual's intense personal devotion of time and labor to his birds, that bound their sense of self to the others they found at the pet shop. Who else would understand the pain when "the hawk" had struck, or the joy of breeding a perfect red teager cap and watching her first flight? The more one invested in pigeons, the more those that counted the most were others who kept pigeons, regardless of age or race. Flyers needed each other in order to build a reputation and validate their self-worth. The flyers' social world was distinct in that any status gains that they accumulated in their peer network were nontransferable to other contexts. It was precisely this fact that enhanced the solidarity within the group, whose members sometimes derogatively referred to themselves with a smirk as "dog catchers," "mumblers," or even "lowlifes." In deriding themselves with such labels, they imagined the gaze with which the rest of society viewed them.

Neighborhood Change and Cosmopolitan Ties

It was through mapping the biographical intersections between flyers like Carmine and Panama that I discovered how this primary group arose from neighborhood change. In my conversations with the men of color, in almost every case I found that an Italian man introduced them—or their fathers—to pigeons. Thus, it was through their exposure to ethnic whites upon moving into neighborhoods in transition that the men of color began keeping pigeons. Richie, a 50-year-old Puerto Rican who grew up in Bushwick when it was mostly Italian, recounted, 'I used to hang out in Knickerbocker Park, back in the late 60s; I must have been ten. I used to see all the old Italian men—it was all Italian and Irish back then—playing bocce. I was a curious kid, so I asked a lot of questions.' The men would then retire to their rooftops and fly pigeons, and Richie said that his inquisitiveness was rewarded when some of them let him on their roof. At first, 'I was only there to look and learn.' Then he began doing chores for the men, and in return they gave him pocket change so he could start buying his own birds. Richie noted that the flyers' kids moved out of the neighborhood when they got older. 'It was very rare that a son would follow the father.' Richie added, 'But they really wanted to teach someone. What I remember is they would all drink homemade wine, and they were talking all the time to me about pigeons!'

Richie wistfully recalled the imperfect harmony that he said existed between the Puerto Ricans moving in and the established Italian population. 'It was a different time; people followed the rules. Italians might call me a spic, and I would call them spaghetti, but at least we said hi. OK, mostly it was like, you stick with yours and we keep with ours. But people tolerated one another.' When I pressed him about racial tension, he played it down: 'We beat the crap out of each other sometimes, but there was an underlying *respect*.' Richie indicated that the entrance of blacks actually brought Italians and Puerto Ricans closer together: 'Italians didn't really have a problem with us, but they had problems with blacks. When they became more prosperous and started moving to the suburbs, Italian landlords would rent to Puerto Ricans but not to blacks.' Richie brimmed with nostalgia as he recalled Italians like Sal, Vito, and Tony who abided the endless questions of a Puerto Rican boy. 'It was a fun time. It was a different time.' Today, Richie was part of the ethnic group that inherited this neighborhood from the Italians, and he continued flying nearby in East Williamsburg.

Significantly, Richie's remembrance of Bushwick alludes to the fact that blacks faced stiffer resistance than Puerto Ricans when they moved into predominantly white neighborhoods. This legacy may help explain why there were roughly twice as many Hispanics as blacks among the flyers I knew. Yet the basic contours of Richie's story seemed typical of the men of color. Mikey, a middle-aged Puerto Rican flyer, also recalled the former dominance of Italian flyers on Bushwick's rooftops and said that he learned the craft from them as a boy in exchange for his labor. Panama said that he hung out at a homing pigeon club near Bushwick and endeared himself to the Italian president, Joe Larocco, who showed him the ropes. Carlos, a middle-aged Puerto Rican flyer, said that his family was part of the Hispanic wave that washed over white Williamsburg beginning in the 1960s. He said he began helping out an old Italian man named Sonny as a young child and eventually became his partner. Forty years later, Carlos still flew his birds from the same coop, because Sonny left him the building in his will.

Though I could not find historical records that indicate how common this transmission of flying from whites to men of color was, I did locate a firsthand account of this phenomenon. In *A Fancy for Pigeons*, Jack Kligerman depicts a vibrant group of flyers in Brooklyn's Greenpoint neighborhood in the late 1970s that consisted almost entirely of Italians and a few other ethnic whites. While Kligerman noted that some white flyers resisted the encroachment of Hispanics, he nonetheless found that the youngest regulars at a local pet shop were teenage Puerto Ricans, who were just beginning to learn the craft of pigeon flying at that time from the Italians.

In neighborhoods like Bushwick, which went from almost 100% white in 1950 to 3% white in 2000, Italians could no longer reliably recruit their sons or coethnic peers as "partners" to help them clean the coops or carry bags of feed to the roof. Thus, ethnic white flyers turned to incoming Puerto Rican and black boys to assist them on their roofs. While this arrangement was useful to the older men, the boys were often permitted to keep some of their own pigeons and learned how to fly them. When the old-timers departed or died, the men of color carried on.

For some time I wondered why the Italians did not frame pigeon flying as "ethnic." After all, a known history links the pigeon wars and New York flights to Italians, and the old-timers seemed to desperately cling to any practices that marked them as Italian. But now I understood. Just as neighborhoods like Bushwick no longer "belonged" to Italians, pigeon flying no longer "belonged" to Italians or ethnic whites. Keeping pigeons was now shared with blacks and Hispanics. Whether the Italians intended it or not, their pigeons became a mechanism that brought them into contact with people of a different ethnic and age cohort with whom they were not voluntarily associating before. Based on face-to-face interactions, they had transmitted pigeon keeping to new people and settings. Even after Italians relocated to farther-removed neighborhoods, pigeon keeping bound them to newcomers at the pet shop.

Puerto Rican and black men, who did not have a prior collective history of keeping pigeons in New York, have since passed the practice onto their children and friends. A few of the Puerto Ricans and blacks that I knew indicated that they were receptive to this competitive animal practice partially as a result of their familiarity with cockfighting and dogfighting, respectively. In Joey's pet shop, I could see the result of this process of cultural diffusion. Old-timer Italians readily mixed with middle-aged Puerto Ricans and young black men. As the men solidified associations in the pet shop, when they needed new partners for their coop they often recruited them through their peer group at Joey's, or from their neighborhoods. This recruitment could further solidify diverse ties, such as Carmine's partnership with Charlie. In the spring of 2008, a 12-year-old "Spanish boy"[21]—as Carmine described him—named Orlando started keeping flights in his mother's backyard three blocks away from Carmine in Queens. While the boy's father had a coop in Brooklyn, Orlando gravitated to Carmine's roof on a daily basis. Carmine teased and hugged him, and he gave him advice on the birds; and Orlando volunteered to change the watering cans and clean the coops on occasion when he saw Carmine struggling. As Orlando scanned the birds for strays one evening, Carmine grinned and said softly to me, 'He's gonna make a great pigeon mumbler, boy.'

Importantly, in this setting and through this practice, social and cultural differences did not breed animosity and distrust that would prevent relationships across racial and age cohorts. This was a significant break from what many sociologists would expect when neighborhood racial succession occurs in poor and working-class areas.[22] In *Blue Collar Community*, William Kornblum urges sociologists to focus on "the processes which create community attachments that transcend 'primordial' ties of kinship . . . and ethnic descent." He found that, through engagement in shared labor and union politics in South Chicago steel mills, workers from a myriad of nationalities and races learned "to play down their ethnic attachments in favor of personal and civil ties to neighbors and workmates."[23] In the pet shop and on rooftops, I found a situation in which working-class men who still valued rootedness in community and particularistic orientations nonetheless created a primary group that cut across primordial categories. Sociologists have long recognized that class and occupation are logical starting points for building solidarity that mitigates the exclusionary and fragmenting effect of primordial ties. Yet it seems that few neighborhood studies have examined how locals adapt their place-based ties in response to changing social conditions, and how shared tastes and activities—like pigeon flying—can become the bedrock for a more cosmopolitan "we-feeling."[24]

The differences among the flyers could be comical. One time Fat Charlie pulled up at the pet shop blasting reggaeton music through the large trunk woofers of his SUV. The Latin beats shook the truck's body, and when the doors opened three Italian old-timers—Russ, Butch, and Joe—stumbled out holding their hands over their ears. Butch complained, 'I want some oldies!'

The flyers did not always play down their differences. At times, the white old-timers separated out to a corner of the store. Occasionally, the Puerto Ricans switched to Spanish. At other times, the men made light of their differences. Carmine called Charlie's ramshackle coop a "shitty Puerto Rican coop," and some African American flyers playfully called some of the Italians "Guineas." Outside of friendships formed through pigeons, most of the men's social networks hewed closely to their own racial-ethnic group. In those settings, they sometimes fell back on stereotypes to make sense of other groups. For example, I was privy to a few private conversations among older whites in which some of them referred to "niggers" in complaining about crime. This contradiction indicates that the men could still form voluntary and affective associations with particular members of the racial groups they ridiculed in general.

Pigeons, as representatives of the natural world, seemed—in this instance—to transcend the ethnic categories that could divide the men. A shared commitment to, and investment in, pigeon keeping brought the men together on equal footing.[25] As their birds mingled in the sky, the men mingled on the ground. Whatever racial prejudice or pride the men held, they understood that racial categories were not valid resources for gaining the respect of their peers. Perhaps ironically, the flyers formed alliances and sorted themselves out based largely on the perceived "natural" or immutable characteristics of their pigeons. Whether one was a "flight man" or a "tiplet man" was of far greater consequence than one's race. This bracketing of race and ethnicity was fertile soil for the formation of cosmopolitan ties based around collective activity. For the men, such relations came to be unremarkable features of their social world.[26]

Even if the flyers sometimes stumbled through the process of learning how to interact with different kinds of people, keeping pigeons offered a chance to engage in sociability and garner the group's respect. For instance, in fall of 2006 Sal started bringing a man into the pet shop and introducing him as "Mohammed from Egypt." Mohammed, a gentle and amiable man with olive skin in his early 50s, used to play a type of pigeon-thieving game over 25 years ago in Egypt and was interested in flying pigeons again. Sal was excited at the idea of having a new neighbor who flew birds, and he wanted Mohammed to become a part of the pet shop group. I wondered how the flyers would take to Mohammed, given his ethnicity and the fact that he was not particularly interested in keeping flights or tiplets. Though Mohammed seemed not to notice or mind, a few old-timers called him Saddam (it seemed to be an accident) and others kept referring to him as Saudi Arabian. Over the next year and a half, however, Mohammed became a Sunday fixture. He also slowly built a coop and cobbled together a stock of "fancy" pigeons—including Egyptian swifts—that reminded him of his homeland. Mohammed gained the acceptance, and seemingly the friendship, of the regulars. Some began to scout for birds that he liked at auctions, and most kept up on the progress of his coop and sympathized with his problems convincing his wife to let him have pigeons.

While keeping pigeons brought diverse men together, their relations were lubricated by other commonalities. Almost all of them were born or raised in Brooklyn and Queens. They strongly identified with their neighborhoods and usually ventured into Manhattan only if work required. The men's relations derived primarily from "voluntary associations with geographical referents,"[27] and in these changing areas, through pigeons, some of these provincial ties crossed ethnic boundaries. Almost all of the men came from a blue-collar background, and they pursued working-class occupations and lifestyles. This allowed them to base their campaigns for respect on a mutually understood script that sociologist Sherri Grasmuck calls "tough masculinity."[28]

A Man's World

Akin to street corners, the rooftop and the pet shop was a man's world. Flyers generally said that women did not keep pigeons because the work is too difficult and too dirty, but they worked to mark pigeon keeping as a gendered practice. To be sure, most of the men had families and appeared to be committed to them. Joey found a newer, tougher job to support his new child, and he proudly showed off photos of his growing boy. When I asked him why he thought women were not involved in pigeons, he portrayed it as their choice; but when I asked him if his pigeons ever created conflict with his family, he replied tersely, "My girlfriend knows enough to say nothing about the birds." My own infrequency of interaction with women attests to the way the men divided up their social worlds. It was common to enter the house with the flyers—after work or hanging out at the pet shop—and for them to simply acknowledge their wives or girlfriends as we headed straight for the roof. The women I spoke with grudgingly accepted that they could not stop their men from escaping to the roof or the pet shop. They also made jokes or optimistically declared that the pigeons kept the men out of trouble—a mantra that the men also invoked, reproducing stereotypical notions of "manly flaws."[29] Reflecting the traditional gendered division of labor and *leisure* of the working class, the women conceded and even expected that the men—one way or another—would exercise their privilege to spend hours away from the home "in the comfort of the man's world"; whereas for the women, their role and their place revolved around the domestic sphere.[30]

Pet shop regulars still desired the sort of masculine primary groups that their kind have historically sought, a place where they could find sociability and build a reputation among men with whom they shared experiences. Whether Italian, black, or Puerto Rican, these men once gathered at barbershops, taverns, or general stores on their own or with their fathers. Some of them still frequented such locales. The pet shop was a natural extension of, or replacement for, those places that either vanished or no longer met the needs of men who organized so much of their lives around pigeons. It is this type of gritty, improvised gathering place that a working-class urban male could feel right at home in. Complaining about their wives or girlfriends was a sport, and while flyers "allowed" women in the store, they protected it as a masculine space. They once acted as if their civil rights were violated when one of Joey's female friends placed a sign in the bathroom to lower the toilet seat. In this way, the pet shop fostered the continuation of a place-based male sociability that was a part of each of the men's heritage—though it infused it with a sort of "back-fence cosmopolitanism" by facilitating casual, interethnic ties.[31]

The flyers' mix of ethnic and age cohorts indicates that the bases for local social cohesion need not be as centered on ethnicity, or occupational and workplace ties, as most neighborhood studies of the working class suggest. It reminds us that opportunities to collectively engage in mundane activities can mitigate the social categories that divide communities, and it shows how relations with animals can be a facilitator of new social relations. The flyers' animal practices lent substance to these relations as well, giving rise to a unique social world with its own folk knowledge, social hierarchy, aesthetics, and morality that the men used to define themselves. Socialization into the group enabled flyers to notice previously imperceptible "natural" differences between pigeons, which fostered a new appreciation for colors, patterns, and breeds. These biological markers, in turn, became material that the men used to draw social distinctions that allowed each flyer to stake out a place in the primary group. And, in the face of demographic and bureaucratic forces that were remaking the city's neighborhoods, flyers continually adapted their animal practices and their social relations as a way to navigate their ever-changing urban ecologies.

III

Deep Play

SIX

The Bronx Homing Pigeon Club

*Nature, Nurture, and the Enchantment of
"the Poor Man's Horse Racing"*

THE SKY FELT ELECTRIC as Franco Bianchi waited atop his rooftop loft for his homers to return from a 400-mile marathon. It was October 7, 2006, and the occasion was the 14th Annual Frank Viola Invitational. Considered the Kentucky Derby of East Coast pigeon races, it drew contestants from all five boroughs, north Jersey, and southern Connecticut. Franco had bred "young birds" from his finest stock in the spring, sent them on daily "training tosses" at 5:00 a.m. over the summer and fall, and gradually worked up their stamina by entering them in weekly local club races of 100, 200, 300, and 400 miles. This morning, his birds, along with over 1,500 others, were "liberated" from Cadiz, Ohio. If one of Franco's "thoroughbreds" made it home first, he stood to win $50,000. Second or third place would still net him $10,000 and $6,000, respectively.

It had been a rough racing season. Franco lost many of his best pigeons in week after week of brutal storms, and he had little to show for it. It was shaping up to be a long afternoon. Strong northeast winds meant that the pigeons would be fighting their way home, and adding insult to injury, Franco's beloved Yankees were down 3–0 early in an elimination playoff game against the Detroit Tigers. The weather and the baseball game had Franco on edge. Though physics dictated that it would be near-impossible for the birds to arrive before 5:30 p.m., we were camped on the rooftop at 4:30. "OK, I got my cigarettes, got my coffee, got my phone." His dangling cigarette poked out from his yellowed, bushy gray mustache and slurred his barely audible vestige of an Italian accent. Franco got out a chico, a tiplet with clipped wings that he would toss up to lure his homers back to the loft, and held the docile beacon in his hand as he stroked it. He recovered the radio reception just in time for us to hear that the Yankees were now down 4–0. Franco stared skyward, then barked, "Come on! Where *are* you?!" Just then, a pigeon streaked across the horizon. Franco tensed up, then jumped out of his chair and tossed his chico. But it turned out to just be a "clinker," a street pigeon. The chico fluttered back to the roof. "I should stick to hunting," Franco grumbled, "it's easier." At 67, he wondered if he was getting too old for this. Though we chatted a bit, Franco was in the zone. Whenever a clinker, sparrow, gull, or butterfly darted overhead, he would clutch his chico a little tighter.

At 5:45, Franco heard the dreaded chirp of his Nextel phone: "Somebody got one!" Franco rolled his eyes, "There goes my 50,000 out the window!" He tried to soothe himself. "We'll take second." But his phone rang again at 5:57. "I don't want to answer it!" His friend Porpora excitedly reported, "Franco, we clocked!" Each call that followed was like nails on a chalkboard. Marty clocked at 6:04, Dallacco too; and Porpora got the birds he "nominated" for the $50 and $100 betting pools. Franco barely managed to muster an apathetic "Good for you."

The Yankees scored two runs in the top of the ninth, but it seemed too little too late. Franco became convinced that he too had no prospects, even though a $1,000 prize was guaranteed for every bird that finished better than 150th place. Now he just hoped to get a bird back the same day. The sun was setting. There were only a few precious minutes for a pigeon to come home before it would decide, no matter where it was, to go down for the night. We waited until it was pitch black, but nothing came. "Not a fucking pigeon!" The phone rang, and Franco smirked as he yelled into the receiver, "Call 911! I'm going to jump!" But when he heard that "old man Viola" probably won his own race, he grinned. Everybody loved Frank Viola. He turned off the radio. The Yankees lost 8–3; their season was over. The broadcasters were already talking about what adjustments would need to be made next year, and so was Franco.

Pigeon racing has two seasons. In the spring, fanciers race "old birds." But the headliners are the "young birds," those born that year and tested for the first time in the fall in a series of races with increasing stakes. From 2005 to 2007, I followed the Bronx Homing Pigeon Club through three old- and young-bird seasons. I gathered with fanciers on weeknights when they packed their birds at the club for shipment and socialized over weak coffee, joined them on tense weekends while they awaited their birds' return to the loft, and went along for predawn "training tosses."

The tenacity with which homers fight fatigue and the elements—flying over mountains, against strong winds, and through rainstorms—to return "home" from an unfamiliar place hundreds of miles away, arriving thirsty and famished and sometimes missing feathers from encounters with predators, is a marvel of nature that flummoxes scientists and exhilarates fanciers. Yet despite the homing instinct and the untold hours and dollars the Bronx fanciers spent to craft winners, each week many pigeons failed to return. The men walked a tightrope between hope and despair. Once they packed their birds on the truck, fanciers could only wait and wonder. But risk brought opportunity. Every season, a new stock of young birds offered possibility. Each week was a rebirth, as it could be the lucky week that an unheralded fancier or pigeon beat the odds, winning the fancier a modicum of cash and bragging rights.

Anthropologist Clifford Geertz portrayed cockfighting in Bali as "deep" play because matches enacted a dramatized "simulation of the social matrix." The "deepest" fights were those that activated dormant "village and kingroup rivalries," such that a rooster's owner publicly displayed and symbolically risked his and his kin's community standing when he put his cock in the ring.[1] Though of a different sort, pigeon racing was also a social dramatization, with distinct phases. I found that a large part of what made a pigeon race "deep" and engrossing to fanciers was that it playfully pantomimed the struggle between man and nature.

In sending their birds across state lines, fanciers symbolically entered a liminal phase where their fate was at the mercy of weather, predators, magnetic fields, and other unruly and mysterious phenomena that affect pigeons' homing ability. Even as fanciers competed with each other, a shared sense of vulnerability aroused communal sentiment. Shipping night functioned as a ritual of solidarity, signaling the cessation of the solitary, humdrum, and laborious training phase and providing a sociable forum in which fanciers steeled themselves for the climactic finale and invoked superstitious folkways to try to exert control under great uncertainty. "Once they're in the sky," a fancier told me, "they're in God's hands." So when their birds appeared on the horizon—after traveling hundreds of miles without stopping at speeds up to 60 miles per hour—and obediently dropped down to their lofts, the men experienced a miraculous, magical dénouement. Even if they did not win the race, they won back a handcrafted treasure that they relinquished to a mystical and hostile realm beyond the bounds of their control and perception.

While pigeon racing offered monetary and egoistic rewards, and while fanciers took some pleasure in wielding power over another species, accounts that would reduce the allure of this animal competition to instrumentality or exploitation are facile. The seduction of "the pigeon game" lay in the transcendent thrill of a "loyal" pigeon's Odyssean homecoming and the concomitant "collective effervescence" generated on shipping night.[2] The staged, temporally bound clash between nurture and nature imbued racing with an aura of enchantment.

The Club, the Fanciers, and the Birds

Although Parkchester, the working-class neighborhood encircling the Bronx Club, housed mostly Hispanic and black residents by the 1990s, the club endured as a relic from the days when European émigrés occupied the area's vast tenements. The club, surrounded by assorted corner stores and "Spanish" restaurants, abutted the wide and busy White Plains Road just north of the Cross Bronx Expressway and the elevated 6 train's trusses. People double-parked to run errands, men gathered at a barbershop, and rap and reggaeton blasted from cars with tinted windows.

The club occupied the first floor of an unassuming three-story house with aluminum siding and an adjacent brick garage set back from the road and shielded by a chain-link fence. A pigeon loft crowned the house, and two more lofts perched atop the garage like attics. German and Belgian immigrants founded the club in the 1930s, but a sidewalk etching revealed that it moved into the Parkchester site, which members collectively owned, in 1955. The clubhouse was filled with faded Americana from the club's, and the sport's, glory days. A hand-painted sign promoting a charity race from 1953 sat behind a glass cabinet like prized china. And the walls were decorated with a few tarnished plaques commemorating legendary pigeons, along with antiquated photos of banquet dinners at VFW halls and long deceased fanciers in their work shirts proudly holding aloft their homers. The American flag was a frequent motif. A large flag adorned the ceiling of the garage, and the dozens of wooden shipping crates stacked six high below it were painted red, white, and blue. Indeed, the decor venerated pigeons and the nation, invoking the golden era when homers were "war heroes" as messengers for the American army.

A Formica table and mismatched chairs crowded the clubhouse's main

The main room of the clubhouse.

room, which facilitated sociability. Other important items were the overworked coffee machine and the television, which was invariably tuned to the Yankees on game day. A sign on the door to the cramped bathroom read "4U2P," and a chalkboard denoted which fanciers had entered birds in upcoming races. The clubhouse's smaller room was where business was handled. Next to the club president's cluttered desk, a computer stood ready to download the times from fanciers' electronic clocks. A sturdy safe protected the racing dues. The detached garage was where pigeons were registered and packed into crates on shipping night. Its fluorescent bulbs, cinderblock walls, and aged steel fans gave the space an industrial feel.

The Bronx Club was an affiliate of the International Federation of American Homing Pigeon Fanciers (IF), one of several governing bodies for the sport. Founded in Philadelphia in 1881, the IF sets by-laws, helps establish local clubs, provides diplomas for winning birds, issues numbered identification bands, and holds an annual convention. Out of perhaps 15,000 fanciers in the United States and several million worldwide, the IF claims 3,500 members.[3]

To be able to race in the club, members' lofts had to be inside the Bronx's official borders. This ensured that all of the fanciers' birds flew roughly the same distance. However, because the pigeons were liberated from points south (such as Pennsylvania), birds homing to the north

Bronx had to fly several miles farther than those homing to the south Bronx (fanciers called this "overfly"). Because the distance from the point of liberation to each loft was different, the winner was determined by dividing the time taken to complete the race by the distance flown. The bird with the highest velocity, measured in yards per minute, won.

There was no single championship race that determined the best pigeon or fancier, but the men got to test themselves against fanciers from over a dozen other clubs in the greater metropolitan area through "combine races" or "derbies." The Bronx Club organized three of the largest and most prestigious combine races, including the World Trade Center Memorial (which donated a portion of proceeds to the New York Police and Fire Widows' and Children's Benefit Fund). The winners of these races, and those who obtained the fastest combined average speed for all of their birds over the course of the season, won trophies, prize money, and perhaps a mention in glossy magazines like *Racing Pigeon Digest*. Though the men enjoyed getting national recognition, they most aspired to be "the King of New York."

Most fanciers in New York represented the social stratum that sociologist Maria Kefalas calls "working-class heroes": union-proud firemen, police officers, contractors, and sanitation workers—many the children of immigrants—who achieved the American dream.[4] Modestly educated, Bronx fanciers scrupulously invested their wages in their children's college tuition and a humble home in the north Bronx or Yonkers. Despite such costly outlays, generous pensions allowed them to retire before their 60th birthday. Even being able to race pigeons was an indicator of economic success. Many fanciers once kept flights or tiplets, but graduated to homers when their paychecks allowed them to afford the racing clocks, club dues, race entry fees, and other expenses. Some fanciers, like Tucker, said they bought a home based on the strategic advantages of its location for racing. In the three years I observed the Bronx Club, there were between 25 and 30 active members—all male. The typical fancier was white and in his late 50s to mid-60s, although the club had five black and three Hispanic members, five members under age 50, and a few old-timers well over 70 who made regular appearances. Club members had their 15 minutes of fame when they were prominently featured in the 2011 cable television miniseries *Taking on Tyson*, which documented the former boxer's foray into pigeon racing.

All pigeons have an innate ability to return home from unfamiliar places. Yet fanciers greatly improved their stamina and homing ability through centuries of breeding, creating homers that can fly nonstop all day and navigate across 500 miles of unknown terrain. Though the exact mechanism of navigation remains a topic of scientific debate, it seems that pigeons can sense the earth's magnetic fields and use the sun as a compass (they loathe flying at night and can detect ultraviolet light). Once they are within a few dozen miles of home, it appears that they can route-find with landmarks and even smell.[5] Since homers were cheap, easy to breed, and needed little space, pigeon racing became popular among the urban proletariat in late-19th-century Europe. Belgian, German, Polish, and British immigrants later imported "the poor man's horse racing" to industrial US cities like New York, Philadelphia, and Chicago. It is thus fitting that the comic-strip character Andy Capp, a quintessential icon of plebeian life, kept homers.

Training

Training was the biggest constant in fanciers' lives. As soon as homers were of age, they were driven to release points and "tossed" in order to learn how to race home. Tosses started at 5 miles but graduated to 100 miles over the weeks leading up to racing season. Tosses were tedious work, requiring fanciers to fight traffic and park on highway shoulders or empty lots in New Jersey—at the break of dawn—just to release a couple birds. But tosses were small gambles that signaled prospective action. They "psyched up" fanciers for the races to come, both the triumphs and the heartache. Fanciers might discover that the young one bred from their best studs was in fact "garbage"; they might find a diamond in the rough; or their best bird might be picked off by a hawk. Training was one of the most powerful tools that fanciers used to try to reduce the uncertainty of racing. Objectively, training reduced uncertainty by getting birds familiar with the last leg of their race route and helping fanciers identify "quality" pigeons. Subjectively, the fancier gained confidence every time he wrestled his birds back from nature, and "his frame of mind brings peace of mind." Training helped reduce a "painful risk to a calculated one,"[6] allowing fanciers to feel they did everything in their power to win.

Few knew the rigors of training like Marty McGuinness, a smiling, blue-eyed, and ruddy-faced Irishman in his late 50s who sported a white goatee and often wore jeans, sneakers, and a baseball cap. When I first met him, his jocular demeanor momentarily dissolved as he pointed to a 14-carat gold pigeon medal hanging conspicuously outside his clothing.

"I don't fool around. I'm dead serious, like a heart attack. This ain't no hobby with me." Five days a week during the racing seasons, Marty woke up before 4:00 a.m., vacuumed and hosed his loft, packed the birds that he planned on racing into his pickup, and drove to New Jersey to toss them. He planned his vacations around racing season, and he had gone as far as Ireland to find studs that would breed champions for his Queen Ann Loft (every fancier named his loft). His efforts seemed to pay off. Marty was one of the most consistent top-ten finishers in his club and in the citywide races.

On a frigid Monday morning in late October, I met Marty at the club at 5:45 a.m. to go along on a toss. Darkness blanketed the streets, steam hung on my breath, and frost glistened on grassy patches. The coffee was on inside the clubhouse when I arrived, and Marty and five other fanciers were deep in conversation about the good old days of rooftop coops. The fanciers had each come with a "basket" (crate) of pigeons that they paid Marty a small fee to take on the truck along with his birds. When I peeked in the door, Bratton, a black retired cop in his mid-40s, exclaimed, 'Damn, for you to be up here at this time is dedication!' I thought the same thing of them. But training was so ingrained in the men that they found their daily routine unremarkable.

As we warmed our hands on the dash of Marty's pickup, I mentioned that his birds were flying well. Marty chuckled, 'I'll tell you one thing, you never know what's gonna happen. You think you got it, but you could be a champ one week and a chump the next.' Though 'glory only lasts a week,' training was the only path to glory. Those who did not train were called donators because they simply threw their money away. Marty recounted how, before he retired from the sanitation department, he would toss his birds after working the third shift. During the years he trained hens and cocks separately, he might drive to a 100-mile release point twice in one day. Marty grinned as he recalled the year he won 19 individual races plus 2 average-speed awards for the seasons, earning him a plaque that chronicled his victories. 'While they were eating supper, I was training. And while they were eating breakfast, I was also training.'

A bit of light finally kissed the horizon. As we drove beyond north Jersey's sprawl, Marty harped on the value of industry and integrity. He recently began a part-time construction job, and he told me how he was assigned an eight-hour shift to cut out ceiling beams by himself. The job, however, took him only five hours. Rather than take credit for a full day, 'I called my supervisor and told him I was going home. *I'm being honest.*'

Marty imported this ethic to racing. As club secretary, he painstakingly followed each of its by-laws and eschewed behavior that could hint at impropriety. A fancier's best asset, he believed, was a sterling reputation. When the owner of a valuable stray homer that landed on Marty's roof told him (by phone) to just release it, he brought it to the club and liberated it in front of us so that no one could accuse him of keeping it if it failed to return to Connecticut. Marty had voluntarily forfeited races in which he got away with infractions, such as racing a pigeon that he later noticed was not in the proper molting phase. There are many reasons why pigeons' wing feathers may be ahead or behind schedule in molting (e.g., perhaps they were sick). But fanciers have been known to "dope" their pigeons in order to hold back or accelerate molting, giving their birds an advantage on race day by having a wing full of grown flights (outer wing feathers). To combat cheating, clubs had rules that dictated how many flights must be fully molted to be eligible for a race.

We stopped at an empty strip mall at 7:15. The moon still hung in the dim sky, and a hawk buzzed us. Shivering, Marty placed a crate with five of his pigeons on the ground. When he opened the hatch, they refused to fly. 'They know the hawk was here. Isn't that amazing?' Eventually, the birds launched and coalesced in the air, circled to orient themselves, and vanished. Marty commented on each bird, telling me whether it was a "solid" racer or a dawdler being given 'one more chance.' After driving ten miles more, we stopped at a J. C. Penney parking lot. Marty let a few birds out one at a time, explaining that these ones needed to learn how to fly separate from the pack. Then he released a bundle of about 20 together. They flew off straight away without circling to reconnoiter, indicating deep familiarity with the route.

We wound our way to the top of a country road that offered fine views of the valley's fall foliage. The area was a popular place to release pigeons. In just a few minutes, we watched two stocks of homers carom off the valley floor and strafe the hillcrest. As Marty freed the pigeons in bunches, he excitedly pointed out his "silver hen" that had recently won both a club race and the Ed Martin combine race. After the flock hooked to the right, Marty cried, 'Look at her! The silver hen took the lead!' He already anticipated breeding next year's young ones from her.

I could see why Marty expressed pride in the silver hen, and even in his mediocre birds. Every pigeon released that day had survived an intense year. It began with the men consulting their breeding charts and mating up their best old birds in the spring. Each baby then received a special

diet of grains, nuts, and herbs and was monitored daily for signs of disease and stress. At one month old, the "squeakers" were placed just outside the loft in a crate every day for a couple weeks so that they could learn their surroundings. It took several more weeks before they made their first flights. As spring stretched into summer, they worked their way up to one hour of daily exercise. At that point, they were taken on their first training toss, starting as close as 5 miles from home. Slowly, the tosses graduated to 10, 20, 30, 40, and 50 miles. As the training intensified, some birds did not make it home. Stragglers were sold to the pet shop. As the young-bird season commenced, races began at 100 miles and, over two months, increased to 400 miles. In some weeks of brutal weather, less than half of the young birds liberated on race day made it home. Though just a half-year old, the homers Marty liberated that crisp morning were already battle-proven veterans, logging hundreds of hours and hundreds of miles of flying.

Marty called the club to tell Tucker that we tossed from the 50-mile point at 8:10. The birds would beat us home. As we pulled in to the club a little after 10:00, Marty pointed out two stocks of tiplets near the clouds: 'That's a beautiful sight.' Though he once flew tiplets and frequented pet shops, Marty said he grew disinterested as the stocks thinned and turned his attention to racing homers. 'Guys with flights and tiplets, they don't have to train them like we do. When we get done our work, we're too tired to go to the pet shop.' This I understood. After saying goodbye and getting on the 6 train, I promptly fell asleep.

Basketing

Shipping night, or basketing, was a communal ceremony that marked the momentary exit from the lonesome and seemingly Sisyphean reality of training. Racing by-laws required that fanciers personally deliver their birds to the club and register them with officials. Save for some new technology, this rite had changed little over the past half century. The "pleasant dusty scent of pigeon feathers"[7] and the smell of coffee and cigarettes wafted among men outfitted in beat-up jeans, T-shirts, flannels, insulated overalls, and trucker caps. Some sported nylon jackets with an embroidered homer on the back flanked by the name of their pigeon loft (e.g., Cisco Loft and 007 Loft). Basketing transpired in the club's garage on a Thursday or a Friday (depending on the race distance), so the nonretired men came directly from work. Most were in no hurry to leave after their birds were loaded into crates, sticking around to endlessly analyze the weather forecast, the details of their race preparations, their pensions, and the New York Yankees.

Basketing also signaled the arrival of the "squaring off" phase of the action. By parting with their pigeons, fanciers committed themselves to the contest and the risk it entailed. This propelled them into a state of liminality. Separated from their birds and exposed to the vicissitudes of the environment, they could only restlessly await the moment of reunion and fret over whether they chose the right birds and training regimen. The solidarity fostered on basketing night fortified fanciers for this liminal phase, and many of their sociable routines doubled as subtle attempts to ward off nature's malevolent forces.[8]

My first visit to the Bronx Club coincided with basketing for a 300-mile old-bird race. As I entered the low-slung garage, Marty cordially greeted me from behind a table. Vinny, an easygoing olive-skinned Italian American in his early 40s (and son of the club's president, Musto), approached the table with his basket of 15 homers and began to remove them one by one. The birds became instantly limp once he grabbed them. Vinny turned each bird so that its head faced toward him, after which he lowered it in his cupped hands to an electronic scanner. He then tugged the bird's foot so that the scanner could read the microchip on its leg band.

Each homer was assigned an ID number, imprinted on a seamless

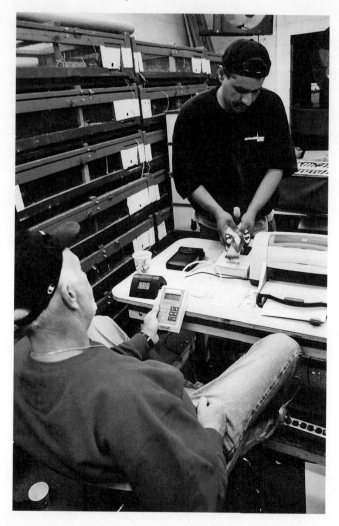

metal band that was placed on the bird's right leg after it was born. For races, fanciers attached a plastic band to the left leg containing a chip programmed with the same number. Marty was responsible for seeing that the number registered by the club's scanner matched the number on the permanent leg band. When the homers returned to their loft, they would alight on a landing pad that "scans them like at the grocery store." The birds' arrival times and band numbers would then appear in the fancier's clock, which he would bring back to the club to have the times recorded by officials.[9]

After each bird was scanned, Oscar, a boyish-looking Puerto Rican man in his 50s, put it in a crate with others of the same sex. As more fanciers came, Musto dutifully sat next to Marty and helped him register birds. An aging, hefty boiler mechanic with oversized glasses and creased jowls, the sullen president barked orders in a booming, gravelly voice that was likely the product of the flavored cigarettes that invariably protruded from his dark mustache. His orders were often aimed at Jonesy, a grizzled, good-humored white man with glazed eyes who seemed run ragged. Too impoverished to race his own birds, Jonesy was Musto's partner. But Musto treated him more like an employee, telling him to make coffee, pack his birds, fill up the baskets with clean water, and so on. By 10:00 p.m., the club filled 14 crates with 25 birds each and loaded them into the ventilated trailer of a large Pullman truck. A hired driver would take the birds to New Stanton, Pennsylvania, to be "liberated" the next morning. Fanciers shipped anywhere from 5 to 35 birds, paying an entry fee of 50 cents each. The winner would receive a nominal cash prize, but some men also placed unofficial side bets by "nominating" pigeons for small wagers of $1, $5, or $10. The first bird to clock for each dollar category would win the sum of the bet money laid down.

Although the night's ostensible purpose was to ship pigeons, one could forget they were there. Due to regular human handling and prior basketing experience, they seemed impassive about being passed over a scanner, placed in a crate with 24 unfamiliar pigeons, and left to sit for hours. Upon leaving the club, they would stay in the crates overnight as the truck barreled down the interstate and parked until it was time for liberation. The homers were like statues as the fanciers handled them, and their compliance enabled the men to almost mindlessly scan and pack them while they directed the bulk of their attention toward ongoing sociable interactions.

Marty claimed that the Bronx fanciers were "the biggest ball breakers in New York." Franco called another flyer a Polack, and when Oscar miscounted the number of pigeons in a crate, Marty joked that Puerto Ricans cannot count. The locus of humor that night was an ad for a children's charity that Oscar had pinned to the wall. It featured a photo of a boy that looked remarkably like a fancier named Bratton, and Oscar wrote next to the boy, "Have you seen my daddy?" As each man entered, Oscar asked, "Who it look like?" Every time someone guessed Bratton, the room erupted in laughter. When Marty found an egg in a crate, he accused Franco of laying it and whispered to me, "If you didn't make fun of him he'd think you were mad at him." Franco simply chuckled and called Marty a "maricón" (faggot).[10] The men built many innuendos around the fact that male pigeons are called cocks. One joked about having a "hot cock" in the race, and another countered that his cock was retired. Yet lewd comments routinely segued into communion over graying hair and fading libidos.

While the "abusive joviality"[11] of the fanciers resembled the rooftop flyers' interactions in Joey's pet shop, open displays of status and bravado were noticeably absent during basketing. Sanctioned races, which produced official winners and losers, rendered reputational competitions moot. Most successful fanciers were content to let the results do the talking. Plus, even less adept fanciers sometimes won first place. This egalitarian ethos was also nourished by a temporarily shared sense of vulnerability. Upon basketing their birds, all fanciers entered a moment of deep suspense and palpable insecurity. Will the winds be favorable? Will a strong storm lead to a "smash" (a race in which most birds are lost)? Though winners and losers would be parsed out in the contest's latter phase, for the moment they were all hostages to nature's whimsy. The sociability of shipping night reinforced feelings of a mutual fate, both in the sky and in life. And the intense moments of solidarity that welled up during basketing served as a symbolic bulwark against chaotic or threatening elements—both natural and societal.

In contrast to the insult rituals, basketing also evinced tender displays of concern and affection. Marty often gloated about his infant grandson, and he openly talked about surviving prostate cancer so that others would get screened. Although they made fun of Musto's weight, the men regularly advised him to improve his diet. They discussed their own surgeries, arthritis, MRIs, and experiences with Viagra; and they conferred about marital problems and how to get their kids into college. Another popular conversational topic was work. They compared the pensions of the NYPD, FDNY, and transit authority; and Marty bugged Vinny to take the test for the sanitation department, calling it "the best job in the city."[12] In helping each other navigate their careers, health problems, and family issues, together fanciers strived to attain the good life.

Basketing was a time to project optimism. The men worked to assure themselves and each other that their efforts would pay off. Vinny admitted he was "not as serious" about racing as others. He stopped training after the first race, and he refused to pay big money for pedigreed birds. "I don't do it for the money," Vinny assured me, but he still had high hopes every week. "Believe it or not, the first race I shipped I won." The win was like the first euphoric high of a drug addict, and Vinny had been chasing that feeling ever since. "That gets you hooked." But it had been three years since he won.

After basketing his birds, Vinny took me up to his loft over the clubhouse. He pointed to his shelves filled with B_{12} pills ("Red Bull for pigeons"), electrolytes, and garlic and elderberry juice. Every week, Vinny tried to find a winning formula. Depending on race conditions and the prior week's results, Vinny would "juice them up" with improvised cocktails. For a long race, he might give them protein-rich peanuts; today, he was trying cane juice. Guffawing, Vinny said that he even once prescribed his birds caffeine pills, "but they don't make the birds win, they make them nervous." He also fiddled with methods to motivate his pigeons to come home faster.

Homers do not know they are racing; they just want to get back to where they came from. The result is a battle of wills between fancier and homer, nurture against nature. Fanciers had to mitigate the birds' herd instinct (a "safety in numbers" defense), which induced homers to stick together upon liberation and slow down to accommodate laggards.[13] One way that fanciers did this was to make use of another basic instinct: reproduction. Vinny, like most fanciers I met, usually raced hens and cocks that were sitting on eggs so that they would dash home to tend their nest (both sexes brood). But Vinny also tinkered with a harder system called "widowhood," in which fanciers separated mated hens and cocks in the loft and only allowed them to see each other just before the race—for "foreplay"—and after the race—for copulation. An aroused cock would rush home to mate with his hen, which Vinny called his "true love" because pigeons mate for life. Vinny joked, "It's just like when men get out of prison; they only want one thing!"

In the three years I followed Vinny, I never saw him win. I expected him to be exasperated that his efforts were for naught, yet Vinny brought the same sanguinity to almost every shipping night. The unpredictability of every race and the unknown factors that affected performance meant that the door to victory was always cracked open. In 2005, even Marty's birds came "late" in most races, and he stuck to the same routines that usually brought success. "It's pigeon racing, and that's what makes it so interesting." Vinny was convinced that if he could line up the right bird with the right weather conditions and herbal remedies, he could notch a victory. Other fanciers encouraged this thinking. Marty made sure to tell me in front of Vinny, "He competes. Every now and then he gets the birds right, and he's right there." One spring shipping night, Vinny held a bird aloft and trumpeted, "This bird got 11th at the World Trade Center Memorial. It came in with no tail, feathers missing—and it was sick!" The implication was that the pigeon could do some serious damage now that it was healthy and competing in a small club race. Later in the year, when Vinny did not clock any birds in a combine, he moaned, "Anybody got a Zoloft?" But he instantly perked up and projected fanciful visions of the future: "Two weeks! I'm putting my best bird in. Since March I set him aside. My best bird." Even when Vinny took a season off, he anticipated success upon his return, hanging around the club and taunting, 'Wait until I come back next year—I'll show 'em how to clock.' Occasional teasing aside, the tacit club norm was to participate in, and validate each other's, wishful thinking.

Seldom did one *fancier* come in first more than two weeks in a row, let alone one *pigeon*. The reward structure was built around this reality, providing enough carrots to keep racers like Vinny motivated. Prizes and respect were given for top 10, and sometimes even beyond top 50, placements. In large combine races, 8th through 150th place all won $1,000. The occasional win had a lasting impact. After Franco won a combine in 2005, he arrived at the club with a new silver Dodge pickup. The truck served as Franco's trophy but also as a reminder to others of the potential rewards that a pound of flesh and feathers, and a bit of luck, could fetch them.

Fanciers hedged their bets by putting up to 35 pigeons in a race and sometimes "nominating" birds for discrete betting pools. This way, one could win money even if he was not in the top ten. If a fancier won some bets and got several birds in 150th place or better in a few combine races, he could more than make his money back for the year and feel content that he was not a "donator." As the 2006 season ended, Marty told me that, even though he had not won a race, he was "ahead" because he "clocked good" most of the season and "got money in all the big races." He added, "That's what it's all about; if you're not winning, then just try to be in the money." As Geertz said of cockfighting in Bali, "the general attitude toward wagering is not any hope of cleaning up . . . but that of the horseplayer's prayer: 'Oh God, please let me break even.'"[14]

Deciding which birds to ship was a careful calculation of risk versus reward. One August shipping night, Marty—who usually flew every week—looked up the weather in several towns along the flight path and decided to skip the race when he saw that heavy rain was likely. His birds were 'molting like crazy,' leaving them with ragged wing tips. Plus, a few of them were "designated" for two upcoming combine races, meaning they wore bands worth $100. Tucker, an African American old-timer who had been racing for over 40 years, was less risk averse. He shipped, but after turning over 20 homers, he put the one in his hand back in his crate, shouting, 'This is a good one! I'm keeping her!' Marty joked, 'What, do you have pets?' Indicating that he sympathized with Tucker's reticence to risk losing a good bird, Marty added, 'I got pets.'

Though fanciers valued winners the most, they often had grudging admiration for birds that repeatedly clocked late but always came back from even the toughest races. Vinny's bird 1984 fit this profile. As he held the bird down to scan its leg one night, he smirked, "1984—he's wild and bouncing all over the place. I should kill him." Vinny yelled at the pigeon as Oscar wrestled it into the crate: "Stay out there!" Smiling, he added, "He'll come home. He always does." Fanciers often gave slow birds a "last chance" by putting them in marathon races. Vinny related, "The weak ones will fly three or four races and then say, 'The heck with you, I'm not coming home anymore.' They become street pigeons. That's how I weed mine out."[15]

Fanciers justified their training regimens and race entrants at length during basketing, and it was customary to praise others' efforts and birds. The men passed around their best pigeons to elicit compliments, such as when Leon said Marty's pigeons "feel like air" and "handle nice." They had a difficult time telling me what "handling nice" meant; it was a gut reaction to a pigeon that just felt good when one cradled it in his hands. Yet all fanciers relied on these judgments in evaluating birds. Did it feel clunky, weak, or heavy? Or did it feel sleek, firm, and light?[16]

Using one's intuition to predict winners was ingrained in the pigeon game. Though many fanciers I knew claimed not to believe in it, one of the sport's central myths is that a pigeon's eyes can reveal whether it is a winner. Believers decode eye color, iris pattern, and pupil shape like astrology charts. I saw fanciers pore over pigeon eyes with jeweler's loupes, and photos of homers for sale in magazines often feature an inset close-up of the eye. There is little objective evidence that "eye sign" and myriad other physical attributes predict winners (Marty's birds that felt "like air" performed poorly). Nonetheless, fanciers habitually invoked such dubious metrics, and basketing was the primary site where this folklore was transmitted and validated. No mere parlor game, the pseudoscience of forecasting winners based on look and feel offered the illusion that one had unlocked nature's secrets and could use them to influence future results. Regardless of the outcome, such a belief functioned as a valuable psychological palliative.

Fanciers obsessively tracked weather predictions on computers or cell phones. As the birds flew hundreds of miles, one could never be sure that they would not hit turbulence along the way. Some websites catered to fanciers, displaying conditions at common liberation points and reporting geomagnetic field disturbances.[17] While weather was obviously important for the race, once one decided on the birds he was willing to risk, there was a sense in which the weather became irrelevant—upon liberation, the birds would have to brave whatever Mother Nature threw at them. But the men did not think this way. Rather, they talked about weather forecasts and possible variations as if their conversations had the power to influence conditions in the sky.

One shipping night, most men agreed when Vinny said, "I heard it was gonna be north by northeast." Unfortunately, this meant that the birds would have to fly straight into the wind. As so often happened after hearing an unfavorable forecast, the men underlined the fallibility of such reports: "It could shift," "It might be like that in the beginning but not at the end," "It might not be a strong wind." If rain was predicted, fanciers pinpointed the percent chance of rain and invoked past races when a higher chance of rain was forecast but sun prevailed. Such talk allowed the men to maintain hope that nature would not defeat them, and to imagine a scenario brighter than the one painted by forecasters. A few men literally prayed for southwest winds and sun. It was only when there were tremendous odds for nasty weather that fanciers became fatalistic. When a hurricane promised to deluge the East Coast, Marty resignedly laughed, "I'll be happy just to get a few home." More than a trite topic of conversation, talking about the weather helped fanciers assure themselves that their birds would be safe from the forces of darkness.

If fanciers had a good feeling about a bird, they might—in addition to betting on it—play the pigeon's band number in the lottery. A good pigeon's ID number became endowed with special powers, compelling fanciers to seek other venues to try their luck. This worked both ways. Lotto players announced during basketing that they played their best bird's number, hoping the gamble would bring luck to the bird. When Musto gambled in Atlantic City or Vinny bet on a horse race, they might narrow their betting choices using similar strategies.

Anna feeds the pigeons in Father Demo Square.

The Basilica di San Marco in Venice.

A seed vendor in Piazza San Marco.

The daily feeding of the pigeons in London's Trafalgar Square.

New York flights. Photos by Marcin Szczepanski.

Carmine watches his birds. Photo by Marcin Szczepanski. Retouching by Peter Herpich.

Abdi inspects his birds in Berlin. Photo by Oliver Hartung.

An ultimatum that many pigeon fanciers told me they had received. Courtesy of Mirrorpix.

Fanciers in the Bronx take a break from inspecting clocks.

Pigeons *by John Sloan. Photograph* © *2012 Museum of Fine Arts, Boston.*

Such gambling practices harmonized with fanciers' folklore about intuiting a winner, their incessant dialogues about weather, and their optimistic rhetoric. These routines operated as *rituals of magic*—"prescribed behaviors in which there is no empirical connection between the means . . . and the desired end." While nothing could be done after basketing to *objectively* affect the race's outcome, fanciers did a lot to *subjectively* create a sense of order in "a world in which players have limited control."[18] Rather than resign oneself to the mystical and perilous forces of nature, fanciers resorted to subtle forms of sorcery aimed at manipulating these forces. Aware of their smallness next to nature's transcendent power, fanciers banded together during basketing and emphasized their shared fate—both through pigeons and their station in life. These ceremonial gatherings generated what Emile Durkheim called "collective effervescence," an emotionally charged "we-feeling" that provisionally banished egoism and competitiveness.[19] Such moments energized and sustained fanciers' belief in racing folklore and equipped them with a symbolic amulet as they prepared for their rooftop sequestration.

Race Day

Unlike horse races, pigeon races are unseen. Fanciers simply ship and receive. Yet race day was no less electrifying to fanciers than fight night is to a boxer. It was a sharp break from "the drab and obsessive routine" and the "thankless preparation" that fanciers performed to arrive at these "all-too-brief appearances in the limelight."[20] On the roof, one "must expose himself to . . . the seconds and minutes ticking off outside his control; he must give himself up to . . . an uncertain outcome." This visceral experience of being the plaything of cosmic forces, of entering a bounded world so "outside the normal round" of life and "full of dramatic risk and opportunity," infused waiting with charm. The higher the stakes (or "fatefulness"), the greater the arousal.[21]

Fanciers stared at the sky and even cried out to their invisible birds, trying to will the miraculous moment when a pigeon would drop from the heavens. Coincidently, these scenes resembled religious iconography in which Christ appears as a white dove descending to his apostles. When a homer finally arrived, the men often found their muscles seized, mouths dry, and hands shaking. Breathing a sigh of relief after corralling their first bird across the landing pad, some fanciers indeed felt that their prayers were answered. They marveled at the "loyalty" and "heart" of their bird, and felt a sense of triumph for having used their own wit and sweat to outmaneuver the elements. But fanciers then entered a new phase of apprehension. Who else clocked? What was their time? Rumors swirled across rooftops, leaving the thirst for resolution unquenched. This unsettled tension could remain for hours, as fanciers had to deliver their clocks to the clubs and wait for officials to tally all of the times before a winner was revealed. But this ambiguity allowed a ray of hope to continue to shine on most every fancier, even as they recalibrated their lofty expectations: "I might make the top ten," "I could still be in the money."

On a sunny spring Saturday, I got to the club at 1:00 p.m. to await the return of Vinny's pigeons from a 300-mile race. Due to fog, the birds were not liberated until 10:15 a.m. Given the wind, fanciers guessed that a good bird would fly 45 miles per hour and arrive at 5:00. Vinny was lounging in the kitchen with Oscar—who helped at the club but did not race—and Tucker—who was killing time before going to his roof. Though he claimed to not be too concerned about today's outcome because it was a small club race for old birds, Vinny regretted that he held back pigeons from the event; the weather was better than expected. The three men played poker in the clubhouse, but they continually craned their necks in case a homer approached. "You get nervous," Vinny laughed. As if on cue, at 2:00 Tucker yelled that he spotted two homers. They all dropped their cards and ran outside, though it was impossible for a bird to fly the race so quickly. They were, of course, clinkers. The card game soon broke up, and Vinny started making trips back and forth to his roof. Tucker went home. "Everybody's on edge," Vinny commented. This cautious mindset was bolstered by mythic tales of woe about times when someone lost a race because his bird came home and he was not up there to chase it into the loft.

Reno called Vinny on his Nextel at around 3:30: "What do you got?" "Nothing," Vinny responded. Vinny remarked to me, "See, he calls me . . . [but] there are always a few guys that don't tell anybody their time, or they lie about their time." I asked why people did this. "They like the suspense. Nobody wants to lose. Every second is crucial to know who is going to win." As I watched some men shave a few minutes off their times when reporting them over the phone, or refuse to share, I realized that the use of mild deceit allowed fanciers to postpone the inevitable letdown that would come from putting their cards on the table. It was a charade they put on for themselves, as much as others, to hold onto hope. One time at the club after a race, Cowboy asked Jonesy how his and Musto's birds did. Jonesy shrugged. "Bullshit." Cowboy yelled, "If you don't want to tell us, say so. But don't lie! Everybody lies!" Bratton noticed that Jonesy had a printout of his birds' results and snatched it from him; the times were not great.

It was usually those who suspected they clocked "late" that bent the truth. As long as the outcome was uncertain, the game was still on. Withholding a mediocre race time might also be seen as an attempt to save face. Yet this merely prolonged the unmasking of the sham; results were publicly posted. Misinformation was so common that there was little stigma attached to lying, and there was almost always some doubt about the winners until near the end.

Vinny ascended to the roof at 4:00 and hid inside his coop so as not to make his landing birds nervous. I observed from the stairwell. He told me how to discern homers from clinkers: "A race pigeon will be way up high, his wings pinned back; he will be diving." But the sense of anticipation blurred perception. Earlier, circling clinkers did not mislead him. Yet now that the time neared in which Vinny could expect the birds, he no longer trusted his intuition. "When you're waiting and waiting, sometimes clinkers start to look like race pigeons." Minutes ticked away as Vinny sat in limbo: 4:10, 4:30, 5:00. He occasionally emerged from the coop to search the sky with binoculars. Nextels crackled: "I got nothing." Vinny threw his hands up. "What the fuck? They should have been here by now!" He quickly dashed back inside—a pigeon just landed and was poking about on the roof. Upon realizing it was a clinker, he shooed it away.

At 5:15, Vinny's phone chirped: "Somebody got one. I don't know who, but the first one is in." A minute after this a pigeon deftly swooped down, unfurling its wings at the last moment to slow itself, and landed directly on the pad. Vinny darted out to make sure it went into the loft, using his giant tennis racket–like "hooples" as prods. He beamed and gave me thumbs up, then immediately took to the phones to relay the information. The lines were buzzing, as others eagerly shared their arrival times: 5:22, 5:14, 5:21. Guys asked, "Where does he fly from?" "How much overfly do I have?" These inquiries were geared toward trying to calculate each bird's yards per minute (ypm), for ypm—not time—decided the winner. Five minutes later, Vinny shouted, "Another one!" It trapped quickly, and Vinny smiled, "He went right in, didn't he? Mine go right in!" He grabbed the phone again to tell others. Though his first bird did not seem to be a winner, and his second bird came five minutes later, sharing was still important. It was a time of lonely waiting on the roof, and getting a bird—any bird—reintegrated the fancier into the group by granting him entry into the phone gossip and making him feel that he was part of the action. The fancier was talked to, and about, by others even if his time was unimpressive.

Pigeon racing lacks an unambiguous moment in which the outcome of the contest is determined. In horse or dog racing, this moment occurs when the finish line is crossed after the last lap. But a pigeon race has as many finish lines as fanciers, and the first bird home may not be the winner. Fanciers thus came to the club ill at ease and eager to join in the group chitchat.

The clubhouse phone started to ring, as fanciers hoped to give or glean information. Vinny asked me to write down the ID number and time of a pigeon: "89804, 17:13:57." This bird, which belonged to Henry Dallacco, seemed to have a fast time. Fanciers began to file in with their clocks after 6:00, sipping coffee, watching a televised horse race, and commiserating about "late" times and lost birds. Because it was just an old-bird club race, the tension was relatively low. Some people left their roof once they got a few birds. "I got my nominator," one old-timer explained, "so what the heck am I waiting around for?" Despite the low stakes, few fanciers would forgo waiting on their roof even though electronic scanners made it possible for the pigeons to clock themselves. Marty emphasized, as others nodded, "I'm up there—that's the whole thrill." A few men with lofts in the north Bronx seemed to have "good birds," and they took out pencils to calculate if they could beat Henry's velocity of 1,285.096 ypm. Reno impatiently asked, "How many miles does Henry have to give me?" These were the last frantic moments of hope. For many, the writing was already on the wall. One shouted, "I am late, late, late!" Another did the math with a calculator and frowned, "No, I can't catch him."

Oscar yelled, "Jonesy! Boss wants coffee!" Musto began plugging fanciers' clocks into the computer and giving them receipts that listed the times and speeds of their birds. With the fog and northerly winds, it was a tough race. Many pigeons were lost. A man whose nominator failed to return lamented, "That's one of my best fucking pigeons." Fanciers speculated about their birds' course home: "They probably went around the rain." I asked Vinny how he did. "Ah, I'm late," he sighed. "I might get money, maybe $20. Something is better than nothing. It's depressing when you're late. You spend all week . . . " Louie "the Lip," a boisterous, older Italian American man, tried to liven Vinny's spirits. "You got them good. On a day like today, you got them good." Indeed, things could have been worse.

Bratton said he only got one bird back, and it was late. "I think I'm going to take up a different sport," he groaned, "maybe dog breeding." And when Cowboy, a jovial white man in his 50s, arrived, he complained that he did not get a bird until about 6:15. But Louie gave Cowboy a harder pill to swallow. Smiling wryly, he pulled out a "checkered white hen" that Cowboy recognized as one of his own. The bird had flown two miles past Cowboy's roof and landed in Louie's loft—at 5:22. Louie turned the knife, "That bird said, Good food over here!" Cowboy was a good sport, laughing. What bothered him more was a "nice blue checkered hen" that failed to return. "I thought she was on the money."

Cowboy began anticipating next week: "I'm gonna be ready. I got mine pumping iron right now. I got the hens doing double time all around the block!" At this, he and Louie slapped hands and chortled. Cowboy's misplaced bird became the story of the night, filling the hours as the men waited until all the clocks were reported and the winner determined. Vinny also brightened: "Next week's another race. I'll fly the same birds, and it's the same distance—so they should be ready." Vinny found some redemption in how his first two birds "dropped," grinning as Oscar commented, "That blue one came in nice! He opened his wings and came right down, and I was like, whoa!" Vinny pointed at me, "You saw it!" He was also content that he got back 21 of 32 pigeons in a tough race. He took extra pride in a homer that came back covered in mud (indicating that it "went down" somewhere), and another that returned missing tail feathers (likely from a hawk attack), because they weathered great adversity to come home.

At that moment, Michael burst into the garage. A gloomy white man in his late 50s with large glasses who was prone to fits of anger, he blurted, "Six seconds! My two sons of bitches circled around the house [before landing], and all I needed was six seconds to beat Henry!" He was enraged, but others laughed when Michael bellowed, "I'm going to kill that motherfucker!" Henry, a mild-mannered Italian American in his 40s, was declared the winner. Excepting Michael, he got handshakes all around. The men studied the copies of the race standings, gave compliments to the top 25 finishers (Vinny got 8th, 17th, and 21st out of 355 birds from 22 lofts), and left to join their families around 9:30 to try to salvage what remained of their Saturday night.

Race day had a basic temporal and affective trajectory. But, compared to club races, the volume was turned up during combines. Earlier races were like a regular season, and combine races were akin to the playoffs. For combines, fanciers in the Bronx Club faced off against people from other clubs whose lofts stretched from north Jersey to southern Connecticut. The stakes, and the emotional intensity, were significantly higher. Combines were marked with an explicit pageantry befitting the rare union of many clubs. For basketing, the Bronx Club rented out the stately Morris Yacht Club on City Island until it burnt down in 2006. Afterward, they moved to the Bronx County Democratic Committee hall in Westchester Square. Though it was a far more modest venue, reminding me of an old VFW hall, the party atmosphere was maintained.

With race entry fees of $100 per bird, organizers splurged on catered food. Sterno cans heated dozens of trays filled with lasagna, meatballs, roast chicken, and broccoli rabe. These shipping nights were like reunions, offering the metropolitan community of 200 or so fanciers a chance to refresh ties. Many men brought their wives and children. Sometimes, a celebrity even showed up. On the eve of the 2006 World Trade Center Memorial (WTCM), the men practically swooned as the lightweight boxer Héctor "Macho" Camacho worked the crowd. Also, signaling the stakes of combines, race officials randomly selected pigeons for drug testing. The birds were placed in cages until they defecated, and the feces were then sent to a lab. Marty and other officials also inspected every pigeon's wing to ensure they were properly molted. Wads of $100 bills were handed to Musto as men cashed their paychecks to enter pigeons into the race.

I spent the 2006 WTCM at Franco's Bronx loft. Though he lived in Yonkers with his daughter, he rented a loft that his friend's father built on 215th Street near the elevated 2 and 5 train, in the working-poor West Indian neighborhood of Williamsbridge (in 2007, he was evicted after the property was sold to a developer). Tucked behind a vacant lot amid a vegetable garden, the giant loft was three stories tall, although each floor was only six feet high. The first floor had all the comforts of home: a television, refrigerator, heater, recliner, and bathroom (with a dirt floor). The second floor was where he kept his breeders as well as retired birds that he grew attached to, such as a 16-year-old one ("He won me a lot of money") and two blind pigeons. He kept his active racing birds on the top floor, which featured a deck from which Franco could sit and observe the sky.

As it was still early, we sipped coffee in the "den." The crusty wood panels were covered with bronze plaques documenting his success flying under the name "Jerome Loft." When a pigeon won a major race, it was customary to have its portrait taken and for it to be given a name. Three framed portraits of winners grabbed my attention. "Angel," I remarked, "1st Joe Parkway Memorial Race, Jerome Loft, Bronx NY, 2004." Franco said that Angel was the name of his dog. "And Carl's Main Event," Franco pointed out. The year after winning the JPM, another plaque showed that Mr. Jerome—the son of the bird that won the JPM—captured the second of the three combine races put on by the Bronx Club. "I need the triple crown. I'm going to win the World Trade!" The WTCM was the best paying, and most prestigious, of the three. Franco chuckled as he described the pigeon depicted in the third plaque, Road Kill. "I had him three years. Never did a fucking thing. But I liked him. He was *cool*. You have a hunch. The third year I sent him to the 400-mile race and he was fifth; I sent him to the 500 and he was fourth. I sent him to the derby [race] and he won! And to top it all off, I found out he couldn't see out of one eye!" The bird earned its name because Franco's friends joked that the meat he brought back from his hunting trips was actually road kill. Franco suddenly jumped up, as if an internal alarm went off. "OK, time to go upstairs. Maybe we'll get lucky and get a bird."

Up on the deck, Franco said, "I got my chair, my chico. And then I wait. Ahh! I relax, take in the sun." But he hardly seemed relaxed. Franco was preoccupied with the fact that the pigeons were not liberated until Monday because of bad weather, meaning they sat in the truck for four days. They would likely return "all busted up." He also worried about the path they would take home from West Virginia. "Some of them will go over the mountains and some will go around them—whoever goes over will win." A storm loomed as well, and Franco wondered if his birds had the moxie to blaze through it. He complained about "getting up at 4:30, five days a week" to train. "But if you don't get up, you don't get the gravy." It was all for this moment, the potential "thrill of winning," which he said flight and tiplet flyers did not experience. Franco coughed as he lit another Marlborough. "I enjoy it. It keeps me occupied. I'm semiretired. What would I do?" Franco's wife passed away, and when he brought his sons to the loft, "they ran away! They say, I ain't going to clean that shit!"

Franco sat in his lawn chair, but thought better of it and got up. Cupping a chico, he stared at the sky as if that might make the birds come faster. Five minutes passed in silence, save for the screeching train and Franco's sporadic cough. The season was wearing on him. "I only got a couple more weeks and then it's over." Seconds became minutes; minutes became an hour. Franco stroked his chico and paced the deck. Word came that two men clocked at 4:50; Franco's forehead creased in worry. "They clocked one in Jersey and one in Queens. We are in deep shit if we don't get one in ten minutes." Because he had overfly on fanciers in Queens and Jersey, he could still win. "Come on! Where are you?" His pleas unrequited, Franco stuck his tongue out. He then lit another cigarette, whistled for the birds, and leaned hard on the railing. "Let's get some money!" Just then, a pigeon alighted on the roof. Franco scurried to scare it into the loft, but it went airborne again. He hurled the chico skyward. It turned out to be a clinker. Franco let out a disappointed exhale. "We had a scare. See what these birds can do to you?" His phone chirped: "Nothing. They clocked 4:48 on the Jersey side." He turned to me, "This pigeon game is funny. You kill yourself and . . . " Three more minutes crept past. "Ugh! Oh well, next year! They're all busted up. People might get one here, one there, but the stock is all busted up." He looked longingly at a passing clinker. "If that was a racing bird, I'd have a good one."

Other fanciers waiting for their birds had similar thoughts. As the minutes ticked slowly by, fanciers began to wonder if anything went wrong or if others had clocked birds but were not telling, and they started to adjust their aspirations. "You got over a hundred prizes. Get a few birds, at least you get your money back." Franco had entered 13 birds and placed bets on some "nominators." After a few more minutes passed, Franco pouted in a sing-song voice, "It doesn't look like I'm going to get my money baaaaaack." "We're in deep shit now. Come on seagull! I need you to be a pigeon!" His phone beeped: "Anything?" Franco winked at me, "Yeah, I've been clocking." After Franco owned up to his joke, the caller asked, "Do we have a chance to beat them?" "Nah. Coupe de Ville got 4:48. The big cash is gone, maybe we'll get the little cash." Franco called out to the heavens, "Bring me my nominator!" He turned to me and griped, "Somebody else must have clocked by now. They only talk to certain people." He added, "Everything is going by here: seagulls, doves, pigeons—except the race birds!"

All of a sudden, Franco froze and then stutteringly whispered, "Here comes one! Do, not, move!" His pleas were heard; the sky coughed up a homer. It dropped so quickly we barely saw it. Hours of inspecting the horizon had made our eyes bleary. Franco trotted to the landing pad and extended the hooples at a snail's pace so as not to scare it, then turned around with his mouth agape. "Oh shit! The clock isn't on!" He ran frantically into the coop, turned the clock on, and waited 30 seconds for it to warm up. He looked like a child that needed to go to the bathroom. Finally, it beeped, and he pushed the bird across the pad. He lost about

The Bronx Homing Pigeon Club 185

Franco recovers his chico after "a scare."

a minute, but looked relieved. "Well, at least I clocked him. Look on the good side. We got one. I hope it's my nominator!" Franco tended to his pigeon like a doting parent, giving the bird its own coop and ensuring that it drank water. He read the band: "WTCM 2. It's my nominator!" He said he had placed bets of $10, $25, and $50 on this bird.

The afternoon was looking up. Franco's whole demeanor changed; his shoulders relaxed. When his phone beeped this time, he smiled. "Did you get one?" Franco tried to sound casual, "Yeah, I got one. About nine after [5:09]. My nominator." Franco set out on an informational fishing expedition, going down his phone list to find out if anyone else got a nominator. "Nine after! I got my nominator! The clock was off! I'm going to stay right here," Franco chuckled into the phone, "I got all night!" As he hung up he said, "That was Harry. If I don't call these fucking guys, they break my balls." Looking like he was about to dance, Franco asked me, "Did you get a good picture of me with the bird?" He was particularly pleased because he said this bird never clocked well before, but that he had a hunch so he nominated it. He finally sat down. "Well, I'm happy. Time for a cigarette. Anything that comes now is gravy." Basking in the afterglow of the adrenaline rush and the feeling that his bird was "in the money," the experience of waiting had been instantly transfigured. Formerly tense and foreboding, it was now leisurely and euphoric. And while Franco felt helpless moments ago, he now felt empowered.

When Franco returned from the bathroom, having left me in charge of clocking (none came),[22] he said he figured that only 20 pigeons dropped, so a lot of prize money was left. An afternoon that once seemed headed for disaster was full of opportunity. "Now I just have to hope that nobody got their nominator." At that moment (5:50), I shouted, "Oh!" Another bird landed. "Come on 14, it's got to be you." It turned out not to be 14, his other nominator, but Franco began whistling, confident that this one was also in the money. When a caller asked how Franco did, he crooned, "I did all right—got two birds. I got my nominator—10, 25, 50—at eight after five. I got the Bronx guys beat." The caller seemed impressed: "Holy shit!" Franco let out a belly laugh and taunted, "Do you need the mattresses?! Don't jump [off your roof]!" The time now seemed to pass quickly, and as it started to get dark around 6:30, two more homers returned. One was his $1–5 nominator. Perhaps he would win that money. "OK," Franco giggled, "whose balls I gotta bust?" He called Dennis. "I got two more!" Dennis sounded annoyed: "Good for you. I got crap." Franco acted unsympathetic. "Are you alive? Are you jumping?"

As the sun vanished, Franco stayed put. He was unlikely to get more birds in the money, but he did not want the thrill of clocking to end. A friend tried to convince him over the phone to leave for the club, to no avail. "I'm going to wait until dark. We might get a couple more. Stick around! What's the hurry?" Like a father, he turned on the porch light to guide his wayward birds home. But no more came. We finally left for the Bronx Club at 7:30 to turn in the clock.

There were not a lot of happy people when we arrived. Bronx fanci-

Fanciers examine the race results.

ers complained that north winds gave guys in Brooklyn, Queens, and Staten Island an advantage. They said their birds had to fly 10–15 miles farther, expending extra energy that dragged down their average speed. Bratton was utterly disgusted. "You don't see no smile on my face, do you? This shit is stressful, man. This will give you ulcers." Rumors were that a hefty, bald black man in his early 40s, who flew as 3 Amigos in Brooklyn, had a bulletproof speed. But it would be hours before this could be confirmed. The small garage was flooded with well over 100 fanciers, many doing back-of-the-envelope calculations, and some speaking in hushed tones.

As race officials calculated the average speed of each fancier's birds, they wrote them on a big green chalkboard that featured the names of all 161 lofts that entered the race. Conversation ceased, and heads turned, whenever a new time was written. As they gathered more information, fanciers continually adjusted their expectations upward or downward. The leader's smile widened as his hold on first place solidified. Some gave him congratulatory handshakes. With his velocity of 1333.103 ypm, those clocking in the 1100s could no longer hope for prizes. Some dejectedly left as soon as this was apparent. One man wistfully eyed the heir apparent and said to no one in particular, "It must feel good." Around 9:30, Marty pushed through the crowd wearing a satisfied grin and put a hand on my shoulder. "I'm still sitting in fifth; now, that could change, but I still got a pretty good bird."

By about 10:15, most concluded that 3 Amigos had won, even though no official announcement had been made. The club was mostly emptied out, save for some diehards and those likely in the top ten who had a big stake in whether they ended up third or fourth. In the end, 3 Amigos won the coveted $50,000 prize. Marty slipped to seventh, but with several birds in the money he wound up well "in the black." Though Franco did not win the "nominator," he got two birds "in the money" (20th and 78th) and was satisfied to get back more than he spent.

The Seduction of Pigeon Racing

As pigeons clocked and the men agonized over their standings on the rooftop, it was all fanciers could do to maintain control of their nerves. During the 2006 Carl's Main Event (CME), Marty's palms sweated and his mouth ran dry. After his bird came in "early," leading him to think he might win, the blood drained from his face and he lapped up water from a hose that his trembling hands could barely hold. Though he had quit smoking, in moments like these Marty relapsed. He said at the club, "I'm a nervous wreck up there. I still get like that. If I lose that feeling, it's time to quit" (he wound up second). Marty remembered how he once banged his head on a post in his haste to clock a bird: "I walked into the club that night with a two-by-four mark on my head!" Tucker recalled a friend who nervously tossed his chico into the screen rather than into the sky. And Vinny mentioned people who threw up while clocking birds. At my incredulous look, Marty added, "Oh yeah. This is what it is all about; that thrill sustains." While waiting for his birds in the 2007 WTCM, Marty reported becoming physically ill and almost accidentally walking off his roof and added, "I'm gonna have a heart attack up there, but boy, what a way to go!" At the club that night, the eventual second-place fancier Anghel was unable to uncurl his shaking fingers; and his voice wavered as he asked others about their time. Cecil Coston, a black retired cop in his early 50s whose loft sat atop the Bronx Club garage, literally wilted under the pressure of waiting to see if his bird would win the 2007 CME combine (it did). His back began to ache, and as the evening went on the pains increased, forcing Cecil to sit down. Even after Cecil got nine birds and the prizes were all likely claimed, he could not leave the roof. The sensual exhilaration of winning could take over the body. Cecil and

Cowboy claimed to "get a hard-on" and give their wives "great sex" after winning a race.

The 2007 CME also shows that it was not just the prospect of winning that was rewarding. Simply gaining back a bird was euphoric. These minor victories—indeed, minor miracles—were themselves addictive features of the action. Even as the combine was declared effectively over based on self-reports, Musto (whose coop was also on top of the club's garage) nervously waited in an office chair he had rolled into the driveway for his bird to come home. He lit his next cigarette with the one in his mouth, cursed the skies, and shuffled to his coop any time a bird came into view. At 6:02, Musto's pigeon landed on the chimney. He choked on his cigarette, then lumbered into the kitchen to grab a can of feed. "I gotta get him!" He shook the can, luring the bird to the landing pad where it scanned. Musto broke into a rare, gleaming smile and spoke to his prize: "I'm glad you're home, baby!" He jubilantly turned to me, "Well, I got my bird!" He became at ease, and was so ecstatic that he went into the kitchen to make the coffee himself instead of ordering Jonesy to do it. Musto seemed as satisfied with his score as Cecil was to get first. Musto had entered only one bird in the race, and so its fate became a very emotionally charged matter to him. Marty was also late but took solace in small victories: "I got 10 of 17 home, including my favorite hen. I said, when she came in, well, now I am happy."

The pigeon game was structured to enable many versions of success. Prizes and esteem extended well beyond top-three placements, and side bets created multiple chances to win. But in the face of birds lost each week, and the eternal toil of training, fanciers greeted the return of *any* bird as a triumph over the elements and the odds—even if no money was awarded. They won back a feathered treasure that they—after lovingly bringing it into this world, caring for it, and training it—willfully surrendered to a realm beyond their control and logical grasp. Though it may be that instinct and careful training were what compelled the birds to return, most fanciers were content to believe on race day that their bird *chose* to "come home." The truly down and out, and defeated and disgusted, were those contestants who failed to get a single bird back.

Sociologist Max Weber depicted the modern era as "characterized by rationalization and intellectualization, and, above all, by the *disenchantment* of the world."[23] Living in a time governed by the cold logic of science and economics, our subjective experience, Weber argued, has been drained of a sense of wonder, mysticism, and transcendence. Through pigeon racing, though, fanciers gained temporary entry into an enchanted world. While fanciers methodically trained their birds so that they would approximate racing machines, the men appreciated that homers were living, fallible creatures that had to navigate around geomagnetic field disturbances, mountains, wind currents, rain, and hawks. This is what made pigeon racing magical: it was full of deep and enduring mysteries, like the homing instinct itself, and the unpredictability of nature.

"We find magic," anthropologist Bronislaw Malinowski noted, "wherever the elements of chance and accident . . . have a wide and extensive range." It is only once "the pursuit is certain, reliable, and well under the control of rational methods" that magic recedes.[24] This is why sorcery thrived among the fanciers. Nature ultimately determined the winner, but fatalism was existentially unbearable. So fanciers attempted to regain control of their destinies by disputing weather forecasts, claiming to be able to intuit a winning bird based on appearance and feel, using their lucky bird's band number as their lottery pick, and even praying. Such folkways were nourished and propagated during basketing, the symbolically freighted time when fanciers who were scattered across rooftops came together to relinquish their birds.

Though the opportunity for cash prizes, and glory, was an attractive aspect of "the pigeon game," the financial stakes were in fact very low in most races. A large part of what sustained fanciers' fascination and made racing deeply meaningful was that the men could engage in a David and Goliath battle with nature. The sport required them to take risks and make themselves vulnerable for a delimited time. But, just as Geertz noted that one's status was only at stake *symbolically* in the Balinese cockfight because the outcome never *really* altered one's social standing in the village, pigeon racing was a playful dramatization of man's struggle to subdue the environment because the men never *directly* exposed themselves to nature's wrath. Pigeon racing was what psychologist Paul Rozin calls "constrained risk."[25] By pantomiming a dangerous situation, it stimulated affective and bodily responses such as fear, adrenaline, and vertigo. But the men could enjoy this sensual experience because they were in fact safe on their roof. Invoking Geertz once more, it was like playing with fire and not getting burned. It was *pigeons* whose lives were actually risked. For these reasons, fanciers heroicized homers. They were imbued with moral character—they "have fight," are "loyal to home," and "show heart."[26] And winners were immortalized on plaques, retired to the breeding coop, and given names.

Through their feathered protagonists, fanciers were vicariously taken in, and taken over, by nature without leaving their roof. By experiencing themselves "as an object controlled by transcendent forces," fanciers "genuinely experience a new and different world."[27] Akin to the experience of standing before a rugged mountain peak or a towering redwood tree, fanciers felt a sense of awe, wonder, and humility as they waited for their pigeons to come home. Yet the charisma of pigeon racing was not about experiencing a sense of communion or kinship with nature. Ever so faintly echoing romantic frontier mythology, the special charm of the "pigeon game" came instead in attempting to tame the wild. The homers, crafted by the men and descended from bloodlines domesticated over a thousand years ago, were the fanciers' western-bound wagons set against the antagonistic backwoods. For men leading utterly domesticated lives in America's largest city, the subplot of this simulated struggle made pigeon racing an otherworldly experience worth organizing their passions, dreams, and lives around.

It was with pride that fanciers called their sport "the poor man's horse racing." Though living in crowded tenements or narrow row houses and laboring in blue-collar jobs, they needed little more than plywood, access to a rooftop, and some gumption to build a "stable"; and their wages were adequate to organize their own Kentucky Derby. In the pigeon, that maligned yet ubiquitous creature so well adapted to the city, fanciers found a felicitous "thoroughbred." These union-proud men took satisfaction in the work required to be a successful fancier, and in the fact that one had to be a jack of all trades—acting as breeder, trainer, "jockey," and veterinarian. As home owners rooted in their neighborhoods, fanciers also appreciated the sport's place-based, communal character. To be a member of the Bronx Club, one had to live in the borough. This meant that men raced against neighbors they had known for decades. The tasks of basketing and turning in one's clock after the race created rites of solidarity that protected fanciers when they were most vulnerable: the first buffered them for the lonely and liminal period of rooftop waiting, and the second marked members' reintegration into the group. Though fanciers competed with each other, in their shared struggle against nature they nourished a community.

SEVEN

South Africa's Million Dollar Pigeon Race

Rationalizing and Globalizing "the Pigeon Game"

A CLOUD OF NERVOUS energy hung over the crowd inside the Superbowl, Sun City Resort's 6,000-seat indoor arena. While "the Las Vegas of southern Africa"[1] has hosted such entertainment icons as Frank Sinatra, Sting, and Rod Stewart, for the moment the star attraction was a group of 2,242 pigeons. The day was February 4, 2006, and the occasion was the Million Dollar Pigeon Race (MDPR). Described by some fanciers as the Olympics of pigeon racing, and offering the largest guaranteed payout of any pigeon race in the world, the tenth annual MDPR attracted some of the world's most avid—and wealthiest—pigeon racing enthusiasts. Fanciers from 28 countries had shipped their most promising young birds to Sun City the previous year, paying $1,000 for every entrant. There, the MDPR's year-round staff of professional loft managers, trainers, and a veterinarian handled all tasks related to raising and conditioning the feathered athletes. Finally, after months of daily training tosses and five warm-up races, the main event was under way. The birds were liberated at 6:00 a.m. and were likely feeling the heat as they traversed the 392-mile desert path back to Sun City. Nearly all of the Superbowl's 5,000 or so spectators had a pecuniary interest in the race's outcome. First-, second-, and third-place finishes promised purses of $200,000, $120,000, and $75,000. And virtually up to the moment that the winning bird arrived at the loft, anyone was free to put a wager on any pigeon.

Fanciers milled about the merchandise stands and quizzed each other about when they expected the birds home. Above them, two massive screens showed a satellite map inscribed with the liberation time and a blinking pigeon icon that marked the birds' likely progress. Waiters served cocktails to the VIP section of the arena while a singer and her pianist did their best to preoccupy the anxious crowd. As it grew nearer to the moment when the increasingly boisterous spectators anticipated the birds' return, the screens went to a live video feed of the loft where the pigeons would land, located about a mile from the Superbowl. At 2:47 p.m., the crowd roared to life. The screen showed a lone pigeon circling the loft. A hush fell over the masses as we intently waited for it to land and cross the threshold that would decode the leg band and reveal its identity. Alas, it turned out to be an ineligible bird that was lost in a training toss.

After another hour passed, fanciers were literally on the edge of their seats. The video montages, music, and special guest appearances could no longer distract the crowd from its insufferable waiting, and the betting counter had emptied out. Then, at 4:04, two pigeons flitted across the video screen and immediately touched down. The crowd erupted. As the hopes and dreams of thousands of fanciers and gamblers hung in the balance, a cruel comedy ensued. Neither bird seemed interested in crossing the threshold that would clock its time, and MDPR rules stipulated that all pigeons had to "trap" by themselves. The pigeons took turns flirting with the threshold, walking up to and alongside of it and then walking away. Though some spectators with money on the line probably fantasized about running over to the loft and chasing the birds in themselves, all they could do was stare at the screen as the pigeons—blissfully unaware that hundreds of thousands of dollars were at stake—preened themselves and caught their breath. Finally, after an agonizing three minutes, one of the birds stalked up to the threshold, cocked its head to look inside of the loft that had served as its home for six months, and slowly walked in.

I first heard of the MDPR by leafing through an issue of *Racing Pigeon Digest* at the Bronx Club. Wim Peters, a renowned South African fancier and veterinarian who authored two books about racing pigeons, had penned an article about his experience at "the world's NUMBER ONE pigeon race."[2] I was intrigued by the event's glitzy aura, and by the unique character of a race in which fanciers ship their birds to another country to be trained by professionals and then return to a single loft. The MDPR also foregrounded the tension, often lurking just below the surface in the Bronx, between the growing centrality of money in the sport and its legacy as "the poor man's horse racing." Whereas the Bronx fanciers once raced solely for certificates and trophies, they were increasingly competing for cash prizes, buying pricey pedigreed "studs" from breeders, and contending with doping scandals. To me, "one-loft races" like the MDPR symbolized the endgame of the trend toward professionalization. After informing Wim Peters by e-mail that I planned to attend the 2006 MDPR, he amiably offered to share his hotel suite with me in Sun City and be my guide. And prior to the weekend of the MDPR, Wim arranged for me to stay with his brother Theo in Pretoria where, for three and a half weeks, I observed and interviewed over a dozen fanciers at their lofts and in a pet shop.

Clifford Geertz famously called the cockfight a story the Balinese "tell themselves about themselves," meaning that the rules and social norms governing matches provided a "metasocial commentary upon the whole matter of assorting human beings into fixed hierarchical ranks."[3] The MDPR was also a "collective text," but what made it an appealing narrative, particularly in a society emerging from the shadow of apartheid, was that it modeled sanguine meritocratic and neoliberal ideals of equal opportunity, free and fair competition, and inclusiveness. Customarily, fanciers could only race against others whose lofts fell within a small geographic area. But the MDPR enabled fanciers from anywhere in the world to compete directly against one another. An Olympic Village supplanted the local club. Fanciers could even participate virtually by simply shipping birds and monitoring race results online. And by guaranteeing that all birds flew the same distance and were housed, trained, medicated, and fed uniformly, the MDPR was said to "level the playing field"—fanciers won or lost based purely on their birds' "true" abilities.

The MDPR's egalitarian patina, however, masked a countervailing story. It is precisely because homemade elixirs, hard work, and expert trapping give "the little guy" a chance to beat the "big spender" (who can simply buy the best birds) that the sport has long been so enticing to people of modest means. By contracting out the myriad tasks of pigeon racing to an array of experts, one-loft races like the MDPR deskill fancying and make it more likely that money will trump skill and knowledge. The $1,000 per bird entry fee also excludes poorer fanciers.

Through the MDPR, I saw how one-loft races radically alter the relation of fanciers to the sport, their birds, and each other. A technical division of labor, designed to prioritize efficiency and uniformity in pursuit of a global market, is laid over what was customarily an individual craft, "making the [fancier] an ever-smaller cog in an ever-larger social organization."[4] On basketing and race day, fanciers were mere observers, having outsourced the handling and trapping of their birds to MDPR staff. And rather than a winning pigeon retiring to its owner's breeding coop, it was sold to the highest bidder because race rules stipulated that all birds were to be auctioned after the race. Thus, in exchange for a promise of competitive parity and huge cash prizes, the MDPR required fanciers to relinquish control, their expertise, and their pigeons. This rationalization of pigeon racing dampens the sport's ethereal aura of magic and loosens the moorings that have anchored pigeon racing in urban communities. Yet while traditionalistic fanciers may interpret the MDPR as evidence that pigeon racing is succumbing to the "disenchantment of the world," to one-loft race enthusiasts it heralds the promise of a "flat world" and brings honor and renewed vitality to a moribund activity.

Pretoria, Sun City, and the Million Dollar Pigeon Race

When I touched down in South Africa, Wim's brother Theo Peters picked me up from downtown Pretoria. With his jovial demeanor, large belly, gray beard, and a booming voice that projected his Afrikaner accent, Theo—60 years old and sporting sandals, shorts, and tinted glasses—struck me as a tropical Santa Claus. A part-time urban-planning consultant, on the ride back to his gated split-level suburban home Theo explained to me—with no apparent resentment—that he and his white colleagues were forced into early retirement after the collapse of apartheid. In the third week of my stay with Theo and his wife, Hillary, Wim arrived with his girlfriend. Though Wim had a full gray beard and a balding pate, he seemed to share few other characteristics with his younger brother. Tall and thin with chin-length hair protruding from the sides of his head, he was notably more reserved and serious. Among the items Wim crammed into his compact car for the 900-plus-mile trip from Cape Town, he brought a dozen pigeons to sell to friends, a box of his books to consign to the local pet shop, and a DVD about pigeons from around the world.

In my time in Pretoria, Theo, Wim, and I socialized with local fanciers at their coops and at a nearby pet shop. Though there was a buzz of excitement about the upcoming MDPR, only a handful of the 40 or so fanciers I spoke to in and around the city entered their own birds in the race. However, over half of them—including Wim—bought a share of someone else's bird (an arrangement called a syndicate), and many of them anticipated making the trip to Sun City.

I found that most fanciers in Pretoria were Afrikaners, a vestige of the sport's history as a Dutch import and of the capital city's past as the locus of apartheid rule. "Coloured" (mixed-race) men formed their own racing clubs in the Western Cape, where they are the racial majority, in the days when they were barred from racing against Whites, but fancying did not catch on among Blacks.[5] Although the fanciers I met in Pretoria were typically wealthier and more educated than their Bronx counterparts, their routines and narratives were not so distinct. It was in Sun City that I discovered a side of racing that scarcely resembled the sport as I had known it.

Sun City, a large resort 70 miles northwest of Pretoria, opened in 1979 when the property was part of the Black "homeland state" of Bophuthatswana. As sociologist and former croupier Jeffrey Sallaz explains, "An illicit casino industry thrived in South Africa's infamous 'homeland states.' Although the puritanical apartheid regime frequently railed against America's 'Sin City,' in fact these casinos were . . . mirror images of their Nevada counterparts."[6] Because the apartheid government had set aside territories as Black "homelands," akin to Native American reservations in the United States, these places were able to offer entertainment (e.g., gambling, nude dancers) that was illegal in South Africa. The desolate drive from Pretoria evidenced the legacy of Black homeland poverty amid Sun City's luxury. Along the way, I passed squatter camps, derelict markets, mule trains, and signs warning drivers to be wary of their surroundings.

The resort, sprawled amid the dense subtropical vegetation of the bushveld and ringed by low green mountains, was opulent. The Disneyesque Lost City Hotel, lined with jungle-themed frescoes and supposedly modeled after a legendary Palace of the Lost City, offered deluxe lodging that exceeded $300 per night per person. The casino, framed by a monument of larger-than-life casino chips cascading from the ceiling, housed 852 slot machines, craps, blackjack, roulette, and VIP gambling rooms. The Valley of the Waves was one of Africa's largest water parks; and the golf courses were designed by Gary Player and outfitted with live crocodiles in the water. The grounds were a fantasy landscape of palm trees, faux-stone animal statues and Romanesque "ruins," fountains, and manmade waterfalls and beaches. Next door, safari tours offered the allure of *real* animals at Pilanesberg National Park.

According to the MDPR director, Zandy Meyer, the inspiration for

The gateway to the Palace of the Lost City.

the race came from an annual golf tournament hosted by Sun City that, when launched in 1981, was the first golf contest to offer a million-dollar payout. Meyer told me that "there's a massive chasm between the reality of the rewards in pigeon racing versus the cost," and that the sport did not offer a race prestigious enough to satiate the most competitive fanciers' desire for acclaim. With financing from "a consortium of Swiss bankers," the race was able to catapult to the top of the emerging global circuit of one-loft races because, unlike other races that based payouts on the amount of revenue taken in (like the Vegas Classic), the MDPR guaranteed a million-dollar payout.

After operating "at a loss" for the first three years, the MDPR has generated solid profits for its investors. As the former MDPR marketing director unabashedly wrote, "This event is a business. Not a club race where they pay out everything that comes in."[7] In 2006, race entry fees brought in over $2 million. The MDPR banked hundreds of thousands more by taking half the amount of each race bird sold at its auction. Merchandise sales and betting fees were another revenue source. But the price of running the MDPR is huge. The event cost about $500,000 to put on, and the MDPR is not simply an event, it is a bureaucracy with over 100 paid workers and volunteers. In 2006, the MDPR employed dozens of "coordinators" to arrange the shipment of birds from 28 countries to Sun City. Ten staffers cared for the birds on site, including a trainer and "the finest avian vet in the world."[8] Other staff roles included race director, public relations officer, systems administrator, transport driver, marketing director, global ambassador, and so on. The MDPR also retained a sports marketing firm and an attorney. The MDPR's total payout to race entrants in 2006 was about $1.3 million. Just over one million of those dollars were up for grabs on race day, spread across the top 250 placements and several bonus prizes; the rest was dispensed in the form of cash prizes and automobiles to winners of five warm-up races, called "hot spots," that took place in the months before the main event.

Training and Caretaking

Fanciers sent their best juvenile birds to Sun City more than a half year before the race. This allowed the birds to imprint on the MDPR lofts as "home" and ensured that they received uniform care. Whereas fanciers normally oversee all aspects of raising and conditioning their birds over the course of a season, after their young birds were shipped to Sun City, this process was divided into discrete tasks and performed by paid specialists. Shipping birds to Sun City was a transfer of ownership. Once race entry fees were paid, fanciers had no further obligations, or access, to their homers. These birds were auctioned after the race, with owners getting half the proceeds.

Once MDPR coordinators got the birds to Sun City, they fell under control of the trainer, Willie Steenkamp, and his eight assistants. From the casino, I took a shuttle bus to the lofts to meet him. A fenced-in area housed four concrete barracks, each covered in a fresh coat of maroon paint, fronted by green wire screens, and topped by a corrugated metal roof. Having been emptied of the pigeons, which were on their way to be "liberated," the lofts were open to the public. Willie, a boyish South African with a tuft of blonde hair and baby-blue eyes, sat down with me after giving a tour in Afrikaans. He said that being trainer was a year-round job. The pigeons began arriving in the middle of May. He and his staff had to quarantine, bathe, and monitor them for the next two months. In August, Willie began taking them on training tosses. Next came the five 200-kilometer "hot spot" races, which took place over the last quarter of the year. In the two months before the MDPR, Willie said he tossed the birds from 80 kilometers every weekday and from 113 kilometers on

Saturdays. After the race, he and his staff organized the auction. Given the time and energy Willie put into the birds, he was heavily invested in their success. For him, the "greatest satisfaction as a trainer" came from a race in which the "pigeons come good."

For the birds to "come good" on race day, they have to be in good health. This task fell to Rob Conradie, a wiry, balding white South African in his late 50s. Once the birds arrived in May, Rob told me, he started commuting weekly from his home and veterinary practice in Johannesburg to Sun City. Upon intake, he inspected each bird and ran tests if necessary. In the rare event that a pigeon died, he did a postmortem. Rob also placed a "sentinel chicken" in each loft during the quarantine period, which he later "bled out" and tested for avian influenza, salmonella, and other diseases. In the half year between the lifting of quarantine and race day, Rob continued to monitor the lofts every week. He was also in charge of the dope testing.

Basketing

Basketing occurred two days before the main event. MDPR staff loaded all of the pigeons at the lofts into crates for their short trip to the Superbowl, where they were ushered into "the cage"—a roughly 30-by-30-foot mesh-covered area that helped keep any escaped bird within arm's reach. Inside, independent auditors began registering every bird from behind ten large tables. As Paul Smith, the MDPR international race coordinator, explained, this was the first time that each pigeon's band number was exposed since its arrival in Sun City—a process that ensured that MDPR staff could not learn a pigeon's identity and single it out for special treatment. Auditors checked the band number against the computer scanner's readout. Afterward, the auditors again covered the birds' bands, stamped their wings as a receipt, and sealed them in white metal crates.

The large, gleaming white transporter truck sat in the middle of the arena's polished floor, awaiting its precious cargo. As a film crew documented the scene, the truck circled around and workers formed a human conveyer belt to move the crates from the stage into the trailer's designated slots. Soon after the 100-plus crates were loaded on, the truck lumbered out of the arena to begin its roughly 400-mile journey to the liberation point at Gariep Dam in the Free State. The entire basketing process took under four hours, wrapping up around 5:15 p.m.

Some hard-core fanciers stopped by to gawk, catch up with friends,

Pigeons are loaded onto the transporter truck.

and order a drink from the makeshift bar; and vendors sold MDPR gear, pigeon medication, and even jewelry under a row of flags (one for every nation that sent pigeons) and a neon sign that read "The Sun City Million Dollar Pigeon Race." Their voices echoed in the cavernous, largely empty arena, which seemed to swallow up the several dozen people working or milling about under the mezzanine.

Though the MDPR basketing formally resembled the shipping nights that I witnessed in New York, the atmosphere was quite different. While basketing in the Bronx was a lively social event, the MDPR basketing attracted only a trickle of fanciers. Here, basketing was a sterile task to be completed expediently. In fact, Paul Smith congratulated the straight-faced professionals for the speed at which they completed the job. Certainly, a large arena could not capture the intimacy of a small club; relative strangers could not be expected to socialize as if they were intimates; and Sun City offered plenty of diversions that could compel people to skip basketing. But the ultimate reasons for such a stark difference lie in the social organization of the MDPR.

Most fundamentally, fanciers were superfluous at the MDPR basketing. They gave up their birds as youngsters the previous year, and MDPR staff meticulously followed guidelines to ensure that no one knew the pigeons' identities. It was paid staff, not fanciers, that delivered the pigeons to the basketing location; there, hired independent auditors handled all aspects of scanning, marking, and crating the birds. While by-laws in traditional races require fanciers to deliver their pigeons and personally oversee basketing, MDPR rules prohibited this arrangement.

From a bureaucratic standpoint, MDPR rules guaranteed equality and impartiality. Yet the rationalization of basketing seemed to have drained it of its ritual social function, reducing it to a purely instrumental affair. A separate Welcome Reception gala did offer pageantry and sociability, but at a price—one had to purchase an MDPR "weekend package" to be able to attend. The reception actually took place the day *after* basketing, because MDPR officials anticipated that many fanciers would not bother to arrive in Sun City for the Thursday basketing.

The Main Event

The Superbowl was already buzzing at 10:00 a.m. on Saturday, as hundreds of spectators perused the merchandise, looked for friends, and checked the details of liberation. The question on everyone's lips was when the birds would arrive. One jumbo screen replayed video footage from that morning's liberation while the other announced that the birds were released at 6:00 into a sunny sky and slight northeast winds.[9]

As the birds were not due home for hours, I sought interviews with fanciers in the Superbowl (I performed 20 in all). Given that I had yet to meet a female fancier, Wim steered me to two Afrikaner women who raced pigeons—Tersia Engelbrecht and Petra Stiglingh. Both women, who were good friends, entered three birds in the MDPR. Tersia, a farmer from the Western Cape in her early 50s with short, dark curly hair, told me that she started with pigeons in 1992 when her husband asked her to help him. "By rearing the babies . . . I became more interested," Tersia recalled. She jokingly added, "Now *I'm* a fancier and he is *my* help." I had always heard that, for females, the most common route to fancying was through their husbands. But Petra, a painter in her 50s with short blonde hair and wire-frame glasses, told me that, after being introduced to pigeon racing by an old friend, she competed on her own for years outside of Port Elizabeth before marrying another fancier. She and her husband each maintained their own pigeon "team" and competed against each other in their local club. Despite the possibility for tension, Petra said the fact that they both raced birds was "a binding factor." Tersia agreed.

Petra and Tersia blamed traditional gender roles for the dearth of female fanciers. Petra reasoned, "It was a man's sport in the first place. The men usually came together—well, they still do—in a club. And they drink, and they come home late." Meanwhile, "the women must stay home with the kids, and they never go on holidays." Tersia added, "By the time you are through with homework [and] cooking, where can you get some energy to be busy with pigeons?" Tersia pointed out other unique concerns for the would-be female fancier: when tossing birds for training, "you must park your vehicle near the lorries on the highway . . . without being afraid of somebody hijacking you—or worse. That's . . . a man's job."

Petra and Tersia believed that female fanciers were more nurturing and less competitive than men, celebrating the "care work" of fancying in a way that most men did not. As a woman, Tersia claimed, "when a pigeon comes from a race later than the others, it's more special to one's heart than the winner, because you are so glad it came back." Both women said that they cried when their homers died and that they refused to cull sick pigeons, leaving the job to their spouse or giving ailing birds to "the help" to eat. But they did not see gender as a disadvantage. Tersia argued, "A woman is . . . a better fancier than a man." She felt that her "motherly love" enabled her to see small changes in her pigeons' health and behavior that her spouse overlooked.

Despite apparent gender distinctions, both women said that many of their closest friends were male fanciers and that men supported their involvement. It did not hurt, the women added, that they happily performed mundane administrative tasks for their clubs and brought homemade food to basketing and race night. Yet Petra and Tersia added they would appreciate the presence of more female fanciers. Petra expressed bewilderment at women's absence: "You try to persuade them, 'Why don't you like pigeons?' Because it's such a fantastic creature that God created." Tersia responded, "They must have a natural will to be in the sport." Petra agreed, "You can't force them." Somewhat paradoxically, even though both women acknowledged that men erected barriers to female participation, they partly explained women's absence from fancying as a result of "natural" disinclination. Regardless, Petra and Tersia expressed particular pride in competing as females on the grand stage of the MDPR, and they saw the presence of a sizable contingent of female fanciers here as evidence that the sport was opening up to women.

Petra had set up a stand in the Superbowl to sell some of her paintings.

Tersia, Henry, Wim, and Petra in the Superbowl.

As I lingered there, her husband, Henry, stopped by. A fit man in his 50s with graying blond hair and wire glasses, he smiled as he mentioned the balance between cooperation and "nice little competition" that he and his wife had maintained for over 20 years. It seemed like a much easier balancing act than the one that he performed as president of the South African National Pigeon Organization. When Henry's tenure began, as a legacy of apartheid there were still two separate national pigeon racing associations: one for Whites and one for everyone else. Henry told me that he oversaw the consolidation of these two organizations, in the face of substantial resistance from both sides. The hardest part, Henry sighed, was "getting people to trust each other." On a local level, he said that the cleavage—and acrimony—between historically White and Coloured clubs was still apparent across the nation. However, Henry now had a more pleasant job traveling around the world as the MDPR's official international ambassador. He reveled in his role, which gave him the chance to befriend "diverse" fanciers, and he proudly credited the MDPR with "putting South Africa on the map."

By 1:00, the seats in the bottom half of the arena had filled. Tuxedoed waiters served lunch to VIPs by the stage, which variously featured interviews with fanciers and live performances of 1970s easy-listening standards by a black female singer and a white male keyboardist. Zandy Meyer, the MDPR director, took the stage to applause and predicted that the best birds would spend ten hours "on the wing," returning around 4:00. Then Rob Conradie, the MDPR vet, and Wim Peters shared the stage for an interview. Rob discussed the birds' excellent condition on race day and their stage of molting. Wim, introduced as a "man well known to all South African fanciers," discussed the importance of vaccinating one's pigeons, before going on to entertain the crowd with stories of one-mile-sprint pigeon races that he witnessed in Indonesia.

While fanciers from around the world sent pigeons to the MDPR, the vast majority of the crowd—based on my own estimates and those of the MDPR staff I spoke to—were white South Africans. Though most on-stage interviews were in English, some exchanges were in Afrikaans. In those instances, the interviewers rarely offered translations. My conversations in the arena indicated that many people did not enter their own birds in the race. It was typical to buy a share of someone's bird as part of a so-called syndicate. Syndicates enabled people who did not have a pedigreed bird, or $1,000 to spare, a chance to be a part of the action. There were plenty of others in the Superbowl who only laid down bets or who came just as spectators with their friends or spouses. Though there were few female fanciers, there were plenty of women.

Win or lose, some fanciers framed the chance to compete in Sun City as the culmination of their racing careers—if not their lives. William "Butch" Engelbrecht, a 60-year-old Afrikaner from a small town called Schweitzer-Reneke, began racing in a local club at the age of nine. He said his "love for the pigeons" sustained him, chuckling, "It's 50 years later and I'm just as mad about the pigeons now as then." Butch told me that he owned a Toyota auto garage for 40 years before retiring in 2004. As he glanced up at the stage lights, he muttered, "It's wonderful for me to think, coming from such a small place, that here I am in Sun City competing against the best in the world." Butch, who entered 14 birds in the MDPR, was tickled to recount that he had once placed in the top ten at Sun City. Though he noted that the MDPR gets tougher every year with "more overseas entrants and better birds," Butch was emphatic that simply being given the chance to enter birds in "the biggest pigeon race in the world" was "a dream come true." He added, "But if you can win this race or do well in this race, then you know you've got the best pigeons in the world . . . and you've beaten the best fanciers in the world."

While Butch felt a sense of accomplishment by playing in the big leagues despite his small-town roots, Celeste viewed the chance to compete in the MDPR through the prism of his experience as an oppressed minority. A slender, 40-year-old man with dark hair and a mustache who looked to me to be of Indian ancestry, Celeste self-identified as Black. "Black, or Coloured," he shrugged, "same thing for me." Celeste, one of the few brown-skinned fanciers in the Superbowl, said he ran an online tourism company in Cape Town. He told me that his pigeon club—one of the biggest "in the whole continent of Africa"—began in Cape Town's infamous former District Six, a vibrant Coloured neighborhood that was deemed a slum by the apartheid regime, which, beginning in the 1960s, forcibly removed 60,000 residents and branded the area Whites-only. Many of these residents wound up in the notorious Cape Flats, a collection of impoverished "townships" designated for non-Whites. Because of apartheid, Coloured fanciers "weren't allowed to race with the whites." Smirking, Celeste added, "And we didn't want to race with them. These were very poor people racing pigeons, but they did it well." During the bleak years under apartheid, Celeste went so far as to claim that pigeon racing kept him sane. He nostalgically reflected, "I could belong to something. Racing pigeons was a way we could experience our identity as people when we were nobody. We had to create our world within this world that is South Africa, and it was actually a happy world."

Despite lingering racial tensions between historically White and Coloured clubs in postapartheid South Africa, which Celeste called "a small replica of what society is going through," he claimed that it no longer bothered him "if a guy is racist." When I asked why, he breathed deep and stared into his teacup: "I've now got my dignity back. I've got my political right back." Now, with a "fair" chance to beat anyone in the world, he called races like the MDPR "a great leveler." And Celeste could dare to dream about making history. "I would love to be the first Black guy ever to win the Million Dollar, and it's a nice dream. It's a reason to wake up." Reflecting on all he had been through to arrive at this moment, he remarked, "I feel so privileged today to have nine pigeons in the race. I used to come here and I wanted to be a part of it and I couldn't afford it, and somehow it happened that now I can put pigeons in without feeling bad about paying $9,000; and the fact that my wife supports me doing it—unbelievable, the way she supports me." Tearing up, he added, "I'm grateful for that, very grateful." Celeste emanated a sense of satisfaction with his station in life, having gotten a foothold in business, and in the MDPR, in newly democratic South Africa. Now, he said, pigeons helped keep him grounded. "Who gets on his knees every day to clean a pigeon's shit? It's the most humblest thing to do."

I returned to the Superbowl after talking with Celeste in a café just in time for a "false alarm." At 2:47, a pigeon landed and calmly walked inside. An on-stage interview was abruptly halted as the camera zoomed in. But the arena erupted in laughter after it was revealed that the bird returned days late from a training toss. As the moment of reckoning drew closer, the crowd became raucous, and the band belted out its rendition of "I Will Survive" with a sense of urgency.

Celeste told me where I could find the race secretary of the Cape Flats Homing Union. Fabian Allay, 33 at the time, had light skin, gray eyes, and a medium build. The owner of a small computer software and hardware supplier, Fabian could perhaps pass as white in the United States, I thought. But he self-identified as Coloured and said he was classified as such under apartheid. Despite the fact that his club was part of a "federation" with 11 other clubs that competed in 23 yearly combines, Fabian said that most fanciers were still partial to their local club. Given the legacy of racial separation carried out under the Group Areas Act,[10] clubs remained sharply segregated. He said that only 10 of the 111 members of his club were white. Cape Flats fanciers, Fabian noted, were also poor—it was common for a fancier to take "a year or two" off from racing because he could not afford it. This was in stark contrast, Fabian said, to the average fancier from the "White Unions" in the southern peninsula.

Fabian recognized that club allegiance could be a cover for racial pride and prejudice. "We're trying hard to have us all fly under one banner—the Whites, the Coloureds—so that when we do race, we don't want to hear that [southern peninsula] put up this time and [Cape Flats] put up that time. If we're all under one banner, then we only have one best time." Fabian seemed to hope that the federation could achieve a small version of the color-blind society that the nation aspired to. "A few years ago," Fabian marveled, "you'd never find me walking with a White person. It's amazing today: you sleep together and you eat together. Some of the Whites here are some of my best friends." Fabian concluded, "I don't hold malice about what happened years ago. I think life should just go on." However, he admitted feeling a special thrill "when you're racing a dominant White group and you're Coloured and you can beat them on a regular basis." In these moments, Fabian could palpably experience postapartheid social progress. Being able to compete against anyone in the world, he said, magnified this sentiment.

As I finished talking to Fabian, the din of the crowd hinted at the birds' imminent arrival. I joined Theo and Wim Peters in the mezzanine, where everyone was staring at the jumbo screens. The image of a metal rooftop in front of bucolic rolling hills was unchanging, yet the promise of a life-altering winning purse kept fanciers' eyes trained on the displays as if a hidden picture—perhaps of their pigeon landing first—would eventually emerge. There was no longer any audible clapping when the musicians finished a song, and they left the stage shortly before 4:00.

Before I could even perceive the two specks on the screen, the crowd burst into shrieks. Two pigeons swooped down from the hills and deftly landed on the loft roof. The time was 4:04. Screams were replaced by silence as we waited to see which one of the blue barred birds would cross the threshold first. After one walked toward it only to turn and walk away, the crowd let loose a collective sigh. MDPR rules forbade the staff to coax the birds in, as fanciers did on their own roofs, so the impatient crowd was at the mercy of the unhurried birds. After a very long minute passed, one of the birds stopped preening itself and strolled right up to the threshold. The crowd worked itself into a frenzy. Yet it proceeded to walk alongside the boundary and peer into the loft without entering. Spectators fervently shouted at the screen, trying to will the bird—which, no doubt, everyone hoped was their own—into the loft. Still, it would not clock. The announcer beckoned, "Come, come, come!" But the homer actually stepped away and began preening again. Suddenly, the second pigeon loped toward the threshold, which prompted the first bird to reapproach the hatch. They both froze on the border, with an $80,000 differential riding on the randomness of which bird decided to enter the loft first. It was too much for some people to bear, as some closed their eyes or turned away. Finally, an excruciating three minutes was brought to a close as one of the homers casually jumped down into the interior of the loft. Shouts and applause filled the arena, and the second bird jumped back as if jolted by the noise a mile away in the Superbowl. Yet the bird entered to more applause 13 seconds later.

For a few precious moments, until the scanner identified the owner of the pigeon and relayed it to the video screens, everyone could imagine that their bird won $200,000. The crowd seemed to communally hold its breath. And then something seemingly miraculous happened: for the first time in MDPR history, a South African won the race. The bird, named Supermans Sting, was actually bred in Belgium. It flew under the South African flag by virtue of being registered by a South African man who formed a syndicate with the breeder. Nonetheless, South African spectators signaled their approval with a deafening roar as the winner, Vyver Krog, danced and jumped in the aisles. Meanwhile, the announcement of the name and nationality of the second-place contestant, a bird named Amala owned and bred by a Chinese man named Tse Ping, went virtually unnoticed. But his $120,000 check was a significant consolation prize.

The winner, a portly and fair-skinned Afrikaner in his 60s with a bushy gray mustache, was overwhelmed by handshakes as he approached the stage. To the delight of the crowd, Vyver delivered his remarks in Afrikaans. He thanked the breeder, his family, and—to enthusiastic claps—South African fanciers. Fourteen minutes later, three more pigeons landed together. They, too, dallied for a bit before clocking. All three birds were German, winning prizes of $75,000, $50,000, and $25,000, respectively. The first UK bird clocked six minutes later and won $20,000, followed by two South African birds (seventh place, $15,000; eighth place, $10,000). Another homer from the UK ($9,000) and the first American bird ($8,000) rounded out the top ten. By 5:06, the top 20 places were locked up. Like many others, Wim looked displeased. He had great expectations for two birds that he flew as a part of a syndicate, but neither of them had clocked. As the prizes diminished, so did the crowd's enthusiasm. Yet chances remained for fanciers to pick up a few thousand dollars, as cash prizes were awarded for the first 250 birds.

As the crowd began to thin, Theo directed me to an ebullient Afrikaner named Gigi, whose bird had just taken eighth place. I asked Gigi, who said he had been "obsessed" with homers for 53 years, if this was his best showing. He let out a belly laugh, "By a long way, the very best!" Despite confessing that he never expected to do so well, he made the right moves to make it possible. Today's eighth-place winner, he said, was bred out of the ninth-place finisher at the 2004 MDPR—which he bought at the auction—and a hen he bought at an auction in Pretoria. "This was the first year I've got them together," Gigi grinned. "It looks like they clicked!"

As Gigi celebrated, the sky blackened. Though only 25 pigeons came in by 5:30, the loft camera was switched off because of the risk of lightning. Yet 76 more pigeons managed to fight their way through an electrical storm and torrential downpours to clock before nightfall. That night, the prize presentation and celebratory dinner—open to VIP guests only—took place in the Royal Ballroom. I was told that the MDPR director gave a short speech before having a representative of Air Sports International,

Spectators in the VIP section watch for the homers' return.

the leading sponsor of the race, present the top ten finishers with their trophies, gold medals, and checks. To much fanfare, the winner was handed an oversized check for $200,000 and draped in the "winner's jacket."

As I left the Superbowl, I reflected on how the MDPR compared to the races I attended in the Bronx. Here, every pigeon returned to Sun City's loft rather than its owner's coop. Instead of scanning the sky from one's roof or yard, fanciers gathered indoors to watch a screen. When the birds arrived, fanciers did not lure them out of the sky by tossing a chico or walk them into the loft with hooples. Having already forfeited their own training and diet regimens by sending their young ones to Sun City the year before, clocking was another moment where fanciers' practical knowledge was inconsequential to the race outcome. Indeed, spectators noted that the fiasco of the first two birds refusing to trap would not have happened in a club race.

For MDPR participants, the "race season" started and ended in a single weekend. And because all birds returned to the same loft, the winner was immediately apparent. Given that fanciers invested a significant amount of money, but not time, in their birds, it seemed fitting that the MDPR provided a quick resolution packaged as a concertlike entertainment experience. I did not meet anyone at Sun City who was happy simply to get his or her birds back. Though this was a common feeling in the Bronx, that response reflected the massive amount of attention that fanciers lavished on every bird over the course of a half year. Having not made that commitment to, and connection with, the birds they entered into the MDPR, fanciers at Sun City were (understandably) more squarely results driven. The simple fact that a pigeon returned was not experienced as compelling or magical, because the fancier did not personally train it to do so, and because it was not the fancier's home that the bird was returning to. Finally, completing the cycle of estrangement from one's birds, the MDPR auctioned them off the next day.

The Auction

The morning after the race, "the cage" was erected again on the Superbowl floor. Inside, the first 100 birds sat side-by-side in metal pens. Attached to each pen was a sheet detailing the bird's pedigree, breeder, performance in training tosses (in graph form), body and eye color, and sex. Potential buyers and gawkers like me were granted a chance to inspect the birds prior to auction. Given their value, the process was surprisingly chaotic; many birds, especially the champion, spent little time in their cages—they were continually passed from handler to handler and probed, poked, prodded, and even smelled as camera flashes went off in their faces.

The auction commenced at 10:00. The fast-talking auctioneer was almost unintelligible, and his "spotters" shouted and pointed as bidders raised their numbered placards. Prices started high and dropped until there was a taker, at which point bids went up in 5,000-rand increments ($822 at the time). Several well-dressed Chinese men, attended to by a cadre of South African assistants, came ready to buy. China is one of the only countries where pigeon racing is gaining in popularity. Though Chairman Mao banned this "capitalist" pursuit, in 2000 China reportedly accounted for almost half of the seven million fanciers allied with the International Homing Pigeons Federation, and wealthy fanciers were eager to augment their bloodlines with storied European pedigrees.[11] Few were surprised when a Chinese man, "#7," bought the first three birds for 75,000 rand ($12,335), 50,000 rand, and 100,000 rand, respectively. Due to the breeder's reputation, the fourth-place pigeon brought the highest bid: 190,000 rand ($31,250).

As the bidding commenced for the fifth-place bird, the auctioneer began: "Bred by K. W. Wuestefeld, ladies and gentlemen. This is the first

Vyver Krog takes a final look at his winning pigeon before it is sold at auction.

cock on this sale." His words became more rapid as he opened the bidding. "What do you say now, it's started! Do I hear 200,000? Da-ba-da, how about 100,000, 50,000 ha-ma-na, no! Do I hear ten, ha-ba-da, ten, do we got 10,000? Yes we do sir!" #7 got into a bidding war with another Chinese man, #17, who in the end won the bird with a 65,000-rand offer. Of the top 15 birds, #7 and #17 accounted for all but two sales. Fanciers speculated that, perhaps, they were buying on behalf of fanciers back home.

After the 100th bird had been bought, the MDPR weekend was officially over. All birds that finished below 100th place would be sold off in the coming months. The announcer thanked the crowd, the sponsors, and MDPR staff. Glossy flyers advertising next year's race and the new slogan, "Much more than a million," were already printed. Hope springs eternal, and I heard several fanciers swearing to do better next year as they grabbed a flyer on their way out the exit.

The Million Dollar Promise

While the MDPR is a for-profit enterprise, it is run by lifetime pigeon enthusiasts who see their role as more than just a job. Staffers proudly claimed that the MDPR is a race that finally offers an amount of money and prestige that is on par with the investment fanciers make in their sport, and it is a rare instance in "the pigeon game" in which all fanciers are truly granted equality of opportunity. With local pigeon clubs on the decline in most of the world, one-loft races like the MDPR were also seen by their often passionate enthusiasts as a shot in the arm of the sport. While many fanciers may balk at $1,000 race entry fees, race organizers believed that such high-stakes races helped maintain, and even grow, the number of "serious" fanciers.

During basketing, I ran into Saville Penkin, the race coordinator for South Africa. Saville rejoiced at the "tremendous camaraderie between fanciers locally and overseas" that the MDPR facilitated, adding, "Meeting people from all around the world, as opposed to just being in your own backyard with fanciers you compete against in your own club, is a part of the sport that I enjoy so much." Saville claimed, "I'll do anything for the sport," and he saw working for the MDPR as a way to help it. He noted, for example, that the MDPR auction, in which record sums of money have been paid, helped "the pigeon game" to be "held in higher esteem than before."

While Saville acknowledged that pigeon racing was increasingly "for the more affluent fancier," he still maintained that, "If you're rich or poor, whether you have a big loft or a small loft . . . you're absolutely equal once the birds are released. Anyone can win." The MDPR, he went on, was to be commended because its structure as a one-loft race made it "very fair," and its ample prize money and global reach finally provided fanciers with commensurate "accolades" and "financial rewards." Saville concluded, "It's brilliant. It's like the Olympics. You can't do better." Even though only a select number of fanciers could participate in such an event, the MDPR, Saville believed, "puts pigeon racing on the map" and therefore benefits the whole sport.

Rob Conradie, the MDPR veterinarian, said that he fretted about the future of the sport. "You haven't got enough youngsters coming through, and . . . the days of doing it as a hobby are coming to an end because . . . it works out very expensive." While he called this trend his "biggest worry," Rob said it was "pleasing . . . that the chaps that are staying in the sport are doing it more professionally." Indeed, all of the MDPR staff members and participants I spoke to framed the professionalization of the sport—which the MDPR personified—as a positive, exciting trend.

In the midst of the race-day chaos of the Superbowl, I sat down with

Paul Smith, who was in charge of overseas marketing at the MDPR's inception. Smith operated at the upper echelon of the sport. He won virtually every major UK competition and countless international events, served as the UK race coordinator for dozens of one-loft races, and made a handsome living selling his stud pigeons, medicine, and other paraphernalia through Regency Lofts. A smiling, blue-eyed, sunburned Brit in his 60s, Paul played the part of MDPR booster with gusto. At the time, he was the supervisor of all 28 race coordinators. He also arranged syndicates, buying pedigreed birds, entering them in the MDPR, and then selling shares of the birds to others.

Smith immediately launched into his talking points. "Many races are *projected*, not *guaranteed*" prize money; they "offer big first prizes but don't pay them out." Paul also credited the MDPR with the invention of the "hot spot races," which give fanciers five chances to win a car before the main event. He listed other "firsts" and "bests" that made the MDPR such a success: a website with real-time results that garnered millions of hits a year; the use of cameras and satellites to track the pigeons in flight and provide spectators with images of the liberation; and the first to use electronic leg bands. Smith emphasized the value added for race winners aside from their cash prize: "The market out there if their pigeon wins the Million Dollar is out of this world." He bragged that even Queen Elizabeth II entered four pigeons from her royal loft,[12] and he quickly added, "Obviously, it's great for South African tourism."

When I asked Paul to consider the sport's future, he rued, "Pigeon racing . . . has been on the decline, not just in one country, in all countries." He attributed the trend to "what youngsters can do in this age," including "sports clubs, videogames, computers, et cetera." Paul sighed, "It's taken something away from what used to be a good sport in the garden." Nonetheless, he too saw a brighter future refracted through the MDPR. "The good side of it is that the one-loft races are on the increase. Because people see that as a way [to] win money that they would never win at home, even if they raced with the top fanciers there for 40 years."

I had a chance to catch up with Zandy Meyer, the MDPR's founder and director, at the lofts in the relatively quiet race-day interlude between the birds' liberation and arrival. A spry man in his 60s with a dollop of dirty-blond hair and a sun-weathered face, Zandy told me that he was a repairman by trade, but always felt that his passion lay with pigeons. Growing up near Johannesburg, he said his father gave him some common (feral) pigeons as a Christmas present in 1952 and that he made a loft for them out of a 44-gallon drum. Having contracted polio as a child, he found pigeon racing attractive because it did not betray his physical handicap. A successful career in business followed Zandy's time as a repairman, and he portrayed it as destiny that he would find a way to merge his business acumen with his fanaticism for pigeon racing.

Zandy framed the MDPR in terms of equal opportunity: "In normal circumstances, there are many influences other than just pure ability that come into play. Specifically, by the very nature of pigeon racing, pigeons home to their own loft. And we all don't stay in the same house or yard." Given that not all lofts are the same distance from the liberation point, birds that have to fly farther may be disadvantaged because fatigue could slow them down over the last few miles. One fancier could also have an advantage because he has a partner that helps him train, can afford better veterinary care for his birds, dopes his birds, and so on. But the MDPR, Zandy said, "gives the opportunity where people from all over the world can actually compete on an absolutely equal footing." "All of the pigeons are trained the same," raised together, and return to the same loft; and Zandy noted a number of "checks and balances," such as concealing the identity of the pigeons, that prevented a bird from "being disadvantaged." The one-loft system, so his thinking went, meant that the bird that won was "truly" the one with the greatest ability.

Zandy argued that many fanciers spend more money to "procure very good pigeons," care for them, and pay club fees and transport costs than the "prize monies that are normally won." More often than not, Zandy said, "you don't make money with pigeons." But the glitzy MDPR, with its promise of a million dollar bills rather than just more paper certificates, "is the event that does actually give them that opportunity." While that chance may be small, fanciers' disappointment in losing at the MDPR was "tempered with the understanding that at least they've participated, they gave it their best shot, and that they got beaten fair and square."

Another common desire that the MDPR fulfilled, said Zandy, was respect. Fanciers crave "prestige, acclaim, and recognition," yet keeping pigeons carries "a stigma." He was pleased that the MDPR brought "a little bit of snooty value" to pigeon racing because "nobody wants to be ashamed of the sport they practice." Everything about the MDPR signaled that fancying could be a highbrow and respectable activity, from the luxury hotels and casino to the lavish VIP banquets, the international participation and media attention, the corporate sponsors, the state-of-the-art lofts and arena, and the hefty amounts of money paid and given out.

The cockfight, Clifford Geertz argued, made manifest the stark and enduring social inequalities that existed in Balinese society at the time. For instance, rules prohibited matches among villagers from different social strata, thus ensuring that a lowly peasant would never have the chance to rub elbows with, or defeat, a chief. The same men who "dominate and define the society," Geertz noted, "dominate and define the sport." Attending cockfights, then, was "a kind of sentimental education" about a traditional, closed society's fixed status hierarchy.[13]

The MDPR offered participants a very different "sentimental education" about society in an era marked by globalization and neoliberal economic policies: the barriers that have long prevented "free and fair" competition are being dismantled, throwing open the doors to a system in which everyone—regardless of status, geography, race, gender, and so on—has an equal chance to vie for a slice of the pie. Zandy and other MDPR boosters reiterated populist sentiments about the benefits of what journalist Thomas Friedman called a "flat world": the traditional, parochial structure of pigeon racing was a "rigged" system, but, similar to the ways that the information-technology revolution made it possible for villages in rural India to compete for market shares with California's Silicon Valley, the MDPR "leveled the playing field" by obliterating geography and subjecting all birds to the same exact constraints.[14] To boot, fanciers could commune with people from around the globe who shared their passion.

The ideal meritocracy, so often found wanting in everyday life, was realized in the one-loft system. The peasant *could* rub elbows with, and defeat, the chief. Though the appeal of this narrative was universal, some South African fanciers seemed to view the MDPR's delivery on the promise of inclusiveness and fairness as evidence that their country was realizing its democratic and egalitarian ambitions. Most poignantly, non-White fanciers like Celeste and Fabian saw in the MDPR a tangible manifestation of a color-blind society. Though these men viscerally recalled the era when they were subordinate to and segregated from Whites, and though their local pigeon racing clubs were still far from integrated, in Sun City they were able to mingle, and compete "on equal footing," with people of any race or nationality. This is why Celeste's ability to enter birds in the MDPR made him so joyful that he wept. While he was aware that racism and discrimination persisted in everyday life, the MDPR gave him a chance to prevail over Whites. It prompted him to reflect on how far he had come from the days of apartheid, and he believed that it was a microcosm of an even brighter future. The MDPR also seemed to transcend class, status, and gender barriers. Butch marveled that a small-town auto mechanic could go toe to toe with the Queen of England's birds, and Tersia and Petra saw their own positive interactions with male fanciers in Sun City, along with the presence of other female fanciers, as proof that pigeon racing had opened its doors to women. In the end, what the MDPR ostensibly offered was all many people would ask for in life: to be given a fair shake and a chance at prosperity, to be judged and rewarded based on ability, and to be respected. And it showed participants that unfettered global competition was a means of securing these ends.

The Poor Man's Horse Racing?

To fulfill the million dollar promise, pigeon racing had to be rationalized. Sociologist Douglas Harper defines rationalization as the process by which "social organization becomes more systematized as the division of labor becomes more elaborate . . . The knowledge embodied in the social organization increases as the working knowledge of individual participants decreases." Harper adds that "the separation of the worker from the *knowledge* that once guided the work" tends to estrange people from their labor.[15] Certainly, the rationalization of pigeon racing has been ongoing for decades, as the latest technology—such as GPS and electronic clocks—has been applied to try to produce more efficient, predictable results that minimize human error. The increasing rationalization of the sport has a strong correspondence to its commercialization. It has only been in recent decades that combine races in New York and elsewhere began offering large monetary rewards. With so much money at stake, fanciers are compelled to adapt all means at their disposal to gain a competitive edge—and eliminate cheating.

On the whole, though, club races remain a craftsmanlike endeavor. As Harper writes of craftwork, "The knowledge embodied in this work forms a unity; whole processes are controlled by a single individual."[16] The fancier manages all aspects of his birds, mating up his best stock, raising the young, medicating and training them, delivering them to the club for basketing, and trapping them on race day. Over a lifetime of performing these tasks, each fancier has developed the full complement of skills and know-how needed to compete in the pigeon game. One-loft races like the MDPR, however, are an entirely different game in which participants are alienated from their birds and from the working knowledge that has traditionally guided fancying.

By farming out the tasks of fancying to experts, one-loft races fragment and deskill the craft. While rationalization may level the playing field in form, in function it puts those who cannot afford pedigreed birds at a greater disadvantage. In standard races, Wim Peters writes, fanciers "are the musicians, and [pigeons] are the instruments which, if played right, will bring forth beautiful music."[17] But in one-loft races, the fancier is not the maestro. Breeding has always been central to the sport, yet fanciers, especially poorer ones, have long taken comfort in the fact that they can compensate for their birds' lack of pedigree with expertise and hard work. Wim considered it the "saving grace" of fancying that the "poor man" could beat the "rich man."

One-loft races make it likelier that money will trump skill. Whereas one customarily had to be a jack-of-all-trades—excelling as breeder, trainer, and doctor—to succeed in pigeon racing, in one-loft races one can win simply by purchasing shares of other people's pigeons. As Wim told me the night after the MDPR, "Just by spending money, people can participate. They don't have to know anything . . . they can just purchase one [pigeon] and have a full go at this race."

Fanciers in Sun City simultaneously lamented that rising costs dissuaded younger generations and celebrated the MDPR as a remedy for the sport's declining fortunes. But this apparent paradox reveals a truth: the MDPR may attract more *participants* to pigeon racing, but it is not clear that it compels more people to become *fanciers*. Many of those present in the Superbowl had as much stake in the sport of pigeon racing as contributors to an office betting pool for the NCAA Final Four have in basketball—they were casual fans seeking a big payoff.

While fanciers have long raced for money, their participation cannot be reduced to it. The conventional racing season slowly builds over a series of small club races where little is at stake. The social meaning of high-stakes combine races results from their relation to earlier club races. They are like the playoffs, a time where fanciers expect to succeed based on the amount of work and practice they put into the regular season. But most fanciers would not forgo the "regular season" for anything. Older fanciers recall a time when there were few monetary rewards in fancying, yet the sport was vibrant. It was the challenge of unlocking the secrets of nature, the chance to work with one's hands and produce a tangible result, and the strong social ties among fanciers—who were neighbors—that sustained involvement. Few fanciers today would disagree with this assessment, including MDPR participants. One of them, Marius Klingbiel, remarked that the delight of fancying is "about the animal that's coming back to you." The intensity of this moment derives from the fact that it encapsulates a history between fancier and bird: "From the day it's born, I give it food and shelter. It gets bigger. I handle it every day. Eventually it goes into the sky. You train it . . . It becomes like your own son or daughter."

The magic of clocking was a mantra of fanciers that I met at the MDPR. This is perhaps ironic, given that they did not clock their own birds at the MDPR and that many of them did not even enter their own pigeons. Yet it seemed to me that their narratives were a product of their sustained involvement in conventional pigeon racing that is centered on the local club and the backyard loft. The MDPR is an exciting and novel form of action that allows fanciers to participate in an event that transcends the boundaries of the local club, and the nation. Fanciers found it edifying to meet others from around the globe and be a part of a professional event. All fanciers value breeding and look for ways to test the results of their efforts, and all fanciers want to win. The MDPR appealed to this competitive penchant and offered a new means for expressing it on a grand scale. For many fanciers, then, the extravagance and cosmopolitanism of one-loft races are a rewarding addition to—or a fun diversion from—their more mundane fancying careers. While they enthusiastically partook in the festivities at Sun City, their words indicate that their love of the sport remained anchored in the craftlike nature of traditional fancying. While participants would vehemently disagree with one critic's assessment that the MDPR was "all event and no sport" and was "all about the money,"[18] it seems clear that one-loft races like the MDPR contradict the bedrock claim of pigeon racing—reproduced ad nauseam in everyday conversations and pigeon racing magazines—that it is "the poor man's horse racing."

Perhaps the least tangible, but still central, aspect of the sport that is lost in translation to one-loft races is the sense of magic and mystery. Fancying has always allowed for the flourishing of personalized training and caretaking strategies grounded in intuition and superstition. Some fanciers swear by potions made of elderberry extract, while others juice their birds with tea. Some deny their birds sex before a big race, while others insist on prerace mating. In one-loft races, however, the doctrinaire logic of the race organization wipes out idiosyncrasy. Like methodical scientists, race officials apply the same treatment to each bird and control for many of the variables that dictate an outcome in order to isolate, as much as possible, the effect of genetics and breeding. This rationalization, inscribed in a technical division of labor, chips away at the mysteries of homing that imbue racing with a sense of wonder; and the alienation of fanciers from their birds diminishes the aura of magic surrounding clocking—which no longer represents a triumph of the individual over nature or a miraculous rooftop homecoming.

There is a sense in which conventional fancying is a cultural anachronism—a skilled craft, passed down from father to son, grounded in neighborhood ties, not easily amenable to timetables, and laden with local knowledge and an appreciation for the supernatural. To race pigeons, one customarily had to be a member of a local club and maintain a loft within a close geographical proximity to other lofts. Fancying was an inherently place-bound activity, reliant on a community of members who age in place. The Bronx fanciers, pining for this fading world even while accepting and at times embracing the professionalization of the sport, objected to one-loft races like the MDPR on these grounds. Cowboy's comments to me upon my return from South Africa echoed other Bronx fanciers' sentiments. 'I'd never do a race like that. You can't even train your own birds, can't wait on your own roof, can't chase em into the coop.' He then proceeded to boast about the time and labor he invested each day into his birds.

The steady decline of local pigeon clubs is apparent, and seems to be largely a product of demographic shifts and changing lifestyles. Recognizing this, some fanciers see one-loft races as a survival strategy. While events like the MDPR may help keep pigeon racing alive, it is not evident that their popularity is breathing new life into the sport as practiced and adored by the Bronx fanciers (or Andy Capp): a skilled craft of the working class, anchored in city clubs. The brilliance of one-loft races lies in unchaining the sport from locales and making it possible for nonfanciers to participate. Such races offer the allure of prestige and an "equal opportunity" to strike it rich, but they further handicap those who cannot afford to buy pedigreed birds and price out poorer fanciers altogether. In this way, the MDPR is a simulation of the problems, not just the promises, of "free" competition in a "flat world": the rich get richer, exacerbating inequality.

When I talked to Marty about one-loft races and the professionalization of fancying, he embodied the tension of a sport caught between two worlds. He coveted pedigreed pigeons from Europe and was proud that his club put on the most prestigious and largest combine races in New York, but he was devastated when he had to expel a lifelong friend from the club over doping allegations. He rolled his eyes when I mentioned the MDPR, but confided that he once sent birds to a one-loft race in California. Seemingly embarrassed, he frowned, 'It was stupid. I just sent my birds and sat at my computer. I didn't feel nothing.' Marty then wistfully glanced at his old pal Musto and a few other Bronx Club members, linking the sensual thrill of clocking on his own roof to the camaraderie that he felt from competing in a neighborhood club.

EIGHT

Conclusion

Changing Ecologies

IN SEPTEMBER 2009, I traveled to New York from Boston, where I was living at the time, to attend the screening of a documentary about the rooftop pigeon flyers called *Above Brooklyn*.[1] The movie was shown alongside other Brooklyn-related short films on the roof of the Old American Can Factory. Wedged between the grungy Gowanus Canal and the elegant brownstones of Brooklyn's Park Slope, the scene was one I had become quite familiar with. After living in an industry-zoned area in Bushwick and frequenting postindustrial neighborhoods like Williamsburg for six years, as I walked to the venue from the F train, I knew that the dimly lit streets and shuttered factories masked a flourishing network of artists and other "hipsters" drawn by spacious "lofts" and by the aura of authenticity radiating from the grit, cracked sidewalks, and prewar manufacturing plants that stubbornly retained their faded company name.[2]

As I passed through the Can Factory's drywalled studios and ascended to the roof, I rubbed shoulders with other 20- and 30-something–year-olds sipping wine. The rooftop offered sweeping nighttime views of downtown Brooklyn, and as I sat in a folding chair, a band culled from Brooklyn's thriving independent music scene took the stage. After the second film, eight of the pigeon flyers showed up. These gruff, middle-aged Hispanic and black men wearing Yankees caps, sweatpants, and worn jackets were hard to miss among the stylish, predominately young and white crowd. I moved to the last row and watched *Above Brooklyn* with them as they excitedly waited for the parts that they were in and cracked jokes. Carmine rhetorically asked the camera, "How do they come home? I don't understand it!" And Big Frank drew laughs when he deadpanned, "My first wife said, it's either me or the birds—so I said sayonara."

After the movie, the flyers scrutinized the skyline. They expressed amazement at all of the new condominium towers that dotted the area, and they pointed out where long-gone flyers once menaced the skies, including the rooftops of nearby former factories. They were living the film. *Above Brooklyn* portrayed the flyers as urban characters whose time had passed and whose neighborhoods were pulling the rug out from under them. And now, as we scanned the airspace, the men marveled at how the real estate boom and the creative class had remade this manufacturing center and replaced pigeon lofts with "luxury lofts." The irony of a factory turned art space, squarely planted in the postindustrial social and economic order, providing a rooftop venue for young bohemians to consume a film about pigeon coops was palpable.

The following year, Manhattan's Lower East Side Tenement Museum held a public screening of a work-in-progress movie about "the disappearing culture of homing pigeon racing in New York City."[3] Much to my delight, the filmmaker had recruited Marty McGuinness from the Bronx Homing Pigeon Club to speak about his experience. In the hour before the event, I toured the museum. A decrepit walkup apartment building that had remained untouched for virtually a century, it offered a glimpse of the early-20th-century world of working-class, immigrant New York. While the docent detailed the overcrowded and unsanitary conditions of the area—now a haven for trendy boutiques and bars—in that time period, she also harped on the dense community ties rooted in ethnicity and kin that made life bearable and meaningful.

Though the docent did not mention it, I knew that tenement dwellers commonly sought sanctuary from their claustrophobic living conditions on the roof, where with a few pennies and some old cartons they could buy pigeons and jury-rig a coop. I recalled a 1910 painting by John Sloan, an artist renowned for depicting the hoi polloi of early-20th-century New York, that portrayed a man leisurely waving a flag at his pigeons atop a brick apartment building while a boy watches from the parapet (see the color gallery for a reproduction). By midcentury, historian Jack Kligerman writes, pigeons were so popular that "there were coops on every third roof in parts of Brooklyn, Manhattan, [and] the Bronx," and "there were over a hundred [pigeon] stores . . . within a twenty-mile radius of midtown Manhattan." Another writer comments, "In the 1950s, it was hard to walk down a street in Bensonhurst or Coney Island without a flock of pigeons twinkling in the sky."[4]

Folklorist Barbara Kirshenblatt-Gimblett notes that "the pigeon game"

Raising pigeons on rooftops, Lower East Side (ca. 1940). Photo by Roy Perry. Courtesy of the Museum of the City of New York.

necessitates "flat rooftops in close proximity, enough pigeon flyers on those rooftops to fly competing flocks, and cooperation from the owner and inhabitants of the buildings, who must be willing to allow rooftops to be used for raising . . . hundreds of birds."[5] By the 2000s, most parts of New York no longer met these criteria. But in the museum that night, Marty regaled the audience with stories of the days when South Bronx tenements seemingly wilted under the weight of numerous coops. He recounted how local Irish boys were eager to learn the craft, "paying dues" by cleaning old-timers' coops and eventually inheriting their father's loft. Pigeon clubs were numerous, serving as community hubs that hosted galas, charity events, and "Ladies' Auxiliary Clubs."[6] Securing access to a roof was usually easy, Marty said. 'We would give the landlord a few bucks, or we would help out by tarring the roof and fixing things up.' Marty closed by eulogizing the recently deceased legendary fancier Frank Viola. 'Everything's changed. It's a dying sport. It really is.'

While the waning of fancying in New York was precipitated by whites' migration to the suburbs, which disrupted its patrilineal transmission, and by the fact that this dirty and laborious craft looks less appealing to today's youth when compared to alternative leisure pursuits such as video games, the final blow to fanciers in the five boroughs may come from their neighbors. In New York, not even the poorest neighborhoods are immune from rising rents and gentrification. Most fanciers I met claimed to have gone their whole lives without receiving a formal complaint from the city. Rooftop coop evictions were rare. However, in the past decade or so, over half the men I knew reported receiving citations or being subjected to unannounced inspections. It seems that neighborhood turnover brought in some wealthier neighbors who preferred formal mechanisms of control, rather than face-to-face negotiation, to deal with perceived nuisances; and citywide hotlines such as 311 made it easy to file anonymous complaints. After 29 years without incident, in 2006 an anonymous call to 311 led to a court order for Marty to remove his backyard and rooftop coops because they were built without a permit. It took four years, many court appearances, and a slew of letters from longtime neighbors attesting to Marty's character and the birds' cleanliness for him to convince the judge to let him keep the lofts. The caveat was that a lien was placed on his home so that he could not sell it until he removed the coops. 'That's fine with me,' Marty smirked. 'The only way I'm going out of that house is in a body bag.'

Holding up citations and photographic evidence, fanciers grumbled that neighbors blamed them when *street* pigeons nested on their air conditioners or defecated on their cars. Complaints were more likely to come from newer neighbors, who sometimes failed to perceive the differences between the men's "stock birds" and feral pigeons. Long-term neighbors, who often had closer ties with fanciers, were more likely to recognize that the pigeons that annoyed them were almost always feral ones, and they knew that fanciers vaccinated their birds and only let them out for short periods. Most flyers did not have permits for their old coops, and many lofts did not meet safety and fire codes. Thus, inspections have brought evictions. While a few fanciers built new coops that were up to code and registered in order to deal with the new reality of neighbor complaints and bureaucratic surveillance, this was prohibitively expensive for most.

The handshake agreements that many flyers relied on to access roof space from neighbors or landlords have broken down as the properties changed hands. And, while in the old days fanciers' claims to rooftops and air rights were informally recognized as valid by their neighbors, who were usually long-term residents bound together by ethnicity and class, new neighbors seem less inclined to legitimate these claims. Events in Williamsburg, Brooklyn, highlight this tension. Most flyers lost their coops as developers bought up properties in the 1990s. But a half-dozen Puerto Ricans were able to maintain lofts on buildings owned by Los Sures, a sympathetic nonprofit dedicated to providing affordable housing to Williamsburg's Latino population. Yet the housing company's support for rooftop coops crumbled as high-rise condos blossomed in the area and rent-controlled tenement units held by long-term Latino residents were converted into market-rate apartments that attracted the middle class. After growing complaints from new neighbors, in 2007 Los Sures instructed its "super" to evict all coops from its units.[7] Today, most people have little chance of keeping pigeons unless they own a building. Needless to say, many New Yorkers, let alone the working class, cannot afford a home unless they were lucky enough to inherit one. Even rundown houses in gritty sections of Bushwick fetch $500,000.

Growing antipathy toward fanciers is hardly unique to New York. In a move that may portend a trend, Chicago passed a law in 2004 that banned pigeon keeping in residential areas.[8] It was sponsored by Alderman Thomas Allen after he received complaints from his constituents in the historically Polish neighborhood of Portage Park about large coops and their associated odors, noises, and feathers. When I visited his office in 2006, Allen told me he felt compelled to ban the coops because they were a health risk and were interfering with residents' ability to "enjoy their backyard." Chicago fanciers offered to limit the number of birds they kept and submit to annual inspections, but the verdict that Allen, his constituents, and the city reached was that this practice no longer had a place in the modern city. Allen opined, "People don't keep horses in their backyard anymore, we don't keep goats or livestock. What urban was a hundred years ago, and what urban is today, are different. You just have to adjust to the times."

Nature Lost?

While the decline of urban pigeon fancying is a story about neighborhood change, it is also—as Alderman Allen indicated—a story about our changing relationships with animals and nature in the city. Today, only vestiges remain of the human-animal relations that once typified working-class urban communities in the United States and Europe, whose yards—to quote from an ethnography of East London in the 1950s—commonly included "hutches for guinea-pigs, lofts for pigeons, and pens for fowls."[9] Over the course of the 20th century, traces of pastoralism were systematically expunged from city streets in accordance with the consolidation of a bourgeois "imaginative geography" that drew a clear-cut distinction between town and country. As Chris Philo demonstrates, the near-obliteration of urban animal husbandry—largely through zoning restrictions—cannot be fully explained by instrumental concerns like public health. Urban elites felt that the presence of livestock animals and their "beastly" habits (e.g., fornication) threatened a sense of order, civility, and decency in public space. And outlawing urban animal husbandry was thought to be a particularly effective means of civilizing the lower classes, whose alleged immorality and indolence were partly attributed to their habit of sharing spaces with animals.[10]

Although the contrast between town and country is an ideal that seldom neatly maps onto reality, environmental scholars argue that this compartmental imaginary has become embedded in the collective conscience of Western societies as an interpretive frame, organizing people's experience of the environment. Through this lens, the city is idealized as an orderly grid where nonhumans are kept under control and boxed into manicured settings such as parks and gardens. Thus, it is not only livestock that are interpreted as taboo in the modern city. What people classify as "pests" or "nuisance animals" are in fact those species of "wildlife" that trespass on sidewalks and colonize human dwellings in spite of efforts to designate these spaces as human-only places. Their unwitting "transgression" of our "spatial expectations" can be existentially unsettling because it is read as "matter out of place."[11]

As discussed in chapter 2, it is through this spatial logic that pigeons have emerged over the last half century as a significant urban problem. Given their origins as cliff dwellers and their history of domestication in cities, compared to many other urban critters, pigeons stand out as possessing a peculiar partiality for stone and concrete. They mate, defecate, live, and die on our sidewalks and ledges rather than in trees or other "green" spaces, making their pollution of the urban spatial order particularly flagrant and offensive. Though it was once so popular to feed street pigeons that urban parks routinely marked off designated feeding areas or even hosted vendors so that pedestrians could buy feed for the birds, a growing intolerance for chaotic, untamed, and dirty nature in the midst of the city has rendered this animal a "homeless" species. As scavengers

Photo by Jason Eppink.

of humans' refuse, pigeons' figurative pollution of sidewalks is solidified through their association with literal pollution—a linkage captured by the moniker "rat with wings."

Concerns over public health and property damage *do* play an important role in pigeons' designation as a nuisance. However, I have argued that the collective desire to remove them from our streets *supersedes* instrumental interests—it is also a moral matter. As mentioned, the label "rat with wings" was born out of a Parks Department commissioner's efforts to associate pigeons with other symbols of disorder and deviance in New York in the 1960s (e.g., litterers, homosexuals, and "winos"). My interest in studying pigeons grew from the conflation of the "pigeon problem" with the "homeless problem" in Father Demo Square, and the mayors of London and Venice both hitched pigeons to a larger "quality of life" narrative in which pigeons were lumped together with rundown parks, congested streets, and petty crime.

While pigeon feces have the potential to damage property, and while park visitors' desire for clean places to sit is reasonable, the health rhetoric so often employed to muster support for culling pigeons, banning feeding, and evicting fanciers is a red herring. I interviewed 15 epidemiologists who specialized in human-animal disease transmission, and all of them said that pigeons were not even on their radars because the risk of infection through casual contact is so low. Epidemiological studies indicate that it is very unlikely that a person can get ill from pigeons unless she literally ingests or breathes in aerosolized feces in an enclosed space.[12] And a high-ranking New York Department of Health official told me that he knew of no likely scenarios in which humans would contract a disease from street pigeons. Despite these facts, it is common for municipal agents to dubiously deploy epidemiological frames to garner support for control and extermination campaigns against pigeons and other harmless "pests."

While the pigeon may be a paradigmatic pest, it seems that in the city almost any animal can be considered pollution unless it is controlled or civilized. "Invasive species" takes on new meaning. It is notable that the label "rat with wings" is increasingly applied to a variety of birds to signal that they are taboo in domestic spaces: geese, whose feces sully the lawns of business parks and golf courses; gulls, which encroach inland to scavenge; and crows and starlings, which travel in packs and cause a racket in urban and suburban neighborhoods. More than a descriptor, the chimeric metaphor is a discursive resource used to justify extermination efforts.[13]

This trend of banishing both domestic and wild animals from urban and suburban zones, and the concomitant generalized human intolerance for species that flout our "imaginative geographies," is precisely the phenomenon that is lamented in the scholarly discourse that I call Nature Lost. Urbanization, it is said, has alienated humans from other life-forms. While animals were once a source of enchantment, inspiration, and even worship, today our natural tone deafness prevents us from finding meaning in the pigeon's coo or the crow's caw. Our only means of knowing and appreciating nonhumans is by dominating or denaturing them.

Or is it? Despite the continued salience of the town/country dichotomy and the Nature Lost narrative, we seem to have entered a historical moment in which American and European cities are being "renatured," or "greened." Perhaps most notably, concerns about environmental sustainability have fostered a groundswell of support for urban agriculture. While one sign of the desire for local and organic food is the growth of farmer's markets and community gardens, many cities have also eased or overturned zoning laws in order to reintroduce animal husbandry. In Chicago, backyard chicken coops have become so popular that one alderwoman's attempt to ban them in 2007 was met with a sizable public backlash and was quickly shot down by City Hall, which, incidentally, had recently installed a beehive on its "green roof" at the mayor's request so that the bees could pollinate flowers. New York City overturned its ban on beehives in 2010 (hundreds of people were allegedly defying the law anyway), and backyard and rooftop chicken coops and beehives have become so popular in the five boroughs that a cottage industry has sprung up to teach and support would-be "urban homesteaders," who tend to be white, middle class, and college educated. These enthusiasts typically frame their animal practices both as part of a sustainable lifestyle and as a means of "bringing nature back in" to the city.[14]

As James Gibson notes, the growing collective desire to bring nature back into the city also manifests in the "consecration" of certain wildlife that settle in our streets. In 2011, two red-tailed hawks built a nest on a 12th-floor window ledge of the library at New York University. Within days, they were christened with the names Bobby and Violet, had their own Facebook page, and became stars of a streaming reality-TV show, thanks to a web camera installed by the *New York Times*. Coming on the heels of New Yorkers' infatuation with the Fifth Avenue hawks known as Pale Male and Lola (the subjects of numerous books, movies, and pilgrimages), Violet and Bobby gathered a devoted following who anxiously watched online and with binoculars as the pair raised a hatchling named Pip on squirrels and rats captured from the park. The so-called hawk cam was one of hundreds of popular web cameras streaming the trials and tribulations of wild raptors to a global audience. Many viewers, including NYU's president, indicated that witnessing up close these birds' efforts to start a family was a rare chance for them to "transcend" their urbanized milieus and commune with "wild nature."[15]

The hawk cam is a tiny example of how cities are embracing, or at least making a bit more room for, wildlife. Many municipalities, for instance, are restoring native habitats such as wetlands, reintroducing local species, and creating "wildlife corridors" to mitigate the effects of habitat fragmentation. And it is now common for cities to require office buildings to turn off their lights at night when birds are migrating, so that they do not crash into the glass. These initiatives are key components of broader national efforts to counteract declining biodiversity.

Akin to the ways that the reconstitution of village life in cities led many sociologists to replace the gloomy Community Lost trope with a redemptive Community Saved thesis,[16] the "greening" of cities has led some environmental writers to supplant the Nature Lost thesis with a discourse that can be called Nature Saved. James Gibson reflects this position in his argument that a "culture of enchantment" is being reawakened in the collective conscience, counteracting the "blasé attitude" so typical of "urbanism as a way of life." Heeding peregrine falcons in the concrete canyons of the metropolis, and even robins at backyard birdfeeders, can bring charm to our sometimes stark, synthetic environments. A walk in the park can recharge our batteries by invoking the bucolic; a community garden can transform a trash-strewn lot into a miniature edible forest; and chicken coops and beehives can give city slickers a chance to experience rurality without leaving home. Indeed, both "urban homesteaders" and city wildlife enthusiasts routinely appeal to pastoral ideals and claim that activities like beekeeping or bird-watching grant them a sanctuary from the hustle and bustle of urban life and fulfill a need to connect with nature. These (re)connections with particular animals and landscapes, Gibson believes, can lead to a "new covenant between society and nature" in which nature "is allowed to exist on its own terms, for its own sake, valuable simply because it is there."[17]

Perhaps. But it seems doubtful that environmental consciousness-raising will dismantle the "sociozoologic" classificatory systems that human groups use to rank and sort species based on the roles we expect them to play in society.[18] While some boundaries that separate humans and animals are being disputed or rewritten in the ways that Gibson celebrates, others seem to be becoming more entrenched. We may revere animals, in their own terms, "out there"—in "wild" places like national parks or other environs where there is little human settlement. But the closer we bring other species into our everyday lives, the more our appreciation for them is contingent on the extent to which we can fit them into particular socially defined roles and places in our otherwise human communities. While we may rejoice when a regal pair of red-tailed hawks nest on a cornice, our tune would likely change if they flocked to our sidewalks by the thousands.

Somewhat ironically, it seems common for "greening" cities to attempt to repatriate fragile species that were lost to urban development while simultaneously ignoring or seeking to evict the pedestrian species that thrive in the built environment. Aside from revealing a cultural bias for "charismatic" and rare species, I believe that this tension also reveals the durability of our belief in—and fetishism of—"pure," asocial nature. Geese that put off migration to beg us for food, raccoons that invade our trash cans, crows that crack open nuts by dropping them into car traffic, and pigeons that nest on air conditioners and eat from human hands appear to us as a "corruption of natural behavior."[19] In adapting their lifestyles to human habits, these so-called synanthropes subvert our romanticized conceptions of authentic "wild."

Just as the creation of new physical and conceptual spaces for city critters benefits some species but not others, it also produces inequitable effects across human groups. For instance, sociologist Sylvie Tissot shows how the creation of a "dog run" by white middle-class users of a public park was a means for them to carve out a space that excluded "undesirable" working-class minorities who either did not have a dog or felt uncomfortable there. A similar dynamic seems to be at work in the urban agriculture movement, where a very idyllic—and bourgeois—notion of animal husbandry prevails. In her study of backyard chicken coops in New York, sociologist Hillary Angelo found that "homesteaders," most of whom were white, middle-class vegetarians that treated their chickens like pets, censured working-poor Latino immigrants for eating their chickens and keeping them in "primitive" structures. By providing material and educational support to chicken-keepers who followed their principles, by refusing to offer guidance on how to dispose of unproductive hens or unwanted roosters, and by pushing to ensure that slaughter remained illegal, "homesteaders" imposed their version of husbandry onto immigrants and other would-be caretakers.[20] The policy of allowing urban chickens but disallowing slaughter, common in a number of cities including Chicago, decreases the birds' economic utility, disadvantaging poorer people who may actually rely on them. It also detracts from the ostensible environmental benefits of keeping chickens.

Although some "homesteaders" receive only negligible edible goods from their bees and chickens and may see them as quasi-pets,[21] they have successfully framed their practices in the moral language of sustainability. Through this lens, the filth, noise, odor, and minimal but potential dangers (e.g., stings) that these animals pose to neighbors are transmogrified into a picture of bucolic bliss. Unfortunately for pigeon fanciers, they do not fit into this picture. The same cities—and even the same neighborhoods—that hold up "homesteaders" (most of whom are middle class and white) as the vanguard of efforts to make cities "green" frame fanciers (most of whom are working class, and many of whom are minorities or immigrants) as out of step with contemporary urban living, even though fanciers' animal practices are similar in many ways to keeping chickens or bees and seem to place no greater burden on their neighbors. The city of Chicago's installation of beehives on the roof of City Hall and its upholding of the right to keep chickens—including noisy roosters—occurred almost before the ink had even dried on the local appeals court's ruling that it was not a "fundamental right" to keep pigeons in the city.[22]

The Social Experience of Animals

The ways in which the "greening" of cities is a selective process, privileging certain animal and human groups while marginalizing others, belies the Nature Saved claim that such efforts are simple manifestations of a general human yearning to commune with nature. Only certain animals (and plants) are defined as valued "natural" objects, and only certain "kinds" of people seek connections with these animals. This tension makes clear that cross-species encounters are always grounded in the social world. While this may be a straightforward social fact, it is at odds with everyday folk understandings of "nature" as well as some environmental scholarship.

Those who trumpet Nature Saved often recapitulate the Nature Lost story that there was once a time in which humans appreciated and worshipped nature because it transcended social life. In this view, totemism expressed a felt sense of literal kinship with animals and fostered a primordial bond with the natural world. But this idea that our relationships with the natural world were somehow "purer" in "traditional" societies, less polluted by social categories, seems more mythical than factual. We can call this the *myth of asocial nature*, which William Cronon and Michael Bell claim is an artifact of contemporary society's longing for a "moral preserve" that will serve as an "antidote to the ills of an overly . . . civilized modern world."[23]

As stated in the introduction, Claude Lévi-Strauss and Emile Durkheim long ago challenged the myth of asocial nature by showing that clans' adoption of totem animals was in fact a means of creating social solidarity and expressing the relationship between self and society. Not only were clans' relationships with animals mediated by social categories, but the sanctity of such bonds stemmed from their capacity to animate social life. While there are certainly important and dramatic ways that urbanization has altered environment-society relations, debunking the asocial-nature myth suggests a much greater continuity between "traditional" and "modern" environment-society relations than the Nature Lost thesis posits.

An important implication of Durkheim's and Lévi-Strauss's work is that one must always pay attention to the ways in which cross-species encounters are patterned by the particular *contexts* in which they are embedded. This has been a major part of this book's agenda. For instance, while the Nature Lost thesis helps us understand why people today are so apt to loathe street pigeons, it fails to explain why people were so enchanted by pigeons in Piazza San Marco. And while the Nature Saved thesis may partly explain growing interest in urban chicken coops and beehives, it fails to explain why this trend is concurrent with a growing antipathy toward rooftop pigeon coops. By tracing the same animal through a variety of settings, I hope to have shown *how* understandings of nonhumans are shaped by place, class, race, gender, and other social phenomena. Pigeons embodied cultural heritage in Venice, became "bums" in London's newly renovated Trafalgar Square, were symbols of ethnicity for Turkish immigrants in Berlin, and were a source of masculine pride for working-class men in New York.

Recognizing that social life always penetrates our relationships with animals also means acknowledging that, despite the intimations of biophilia and Nature Saved, we sometimes seek relations with nonhumans to satisfy social impulses. Rather than fulfilling a desire to connect with nature, keeping pigeons was a way for the New York flyers to connect to a social world and compete for status. Even the Turkish caretakers' totemic-like "kinship" with tumblers, which they said anchored them in nature, was also a means of anchoring them in their homeland and ethnicity. And feeding pigeons in Father Demo Square seemed to be a response to the social situation of being alone—feeding connected loners to sidewalk life. These encounters, I argue, are not less "authentic" than, or antithetical to, those that foster a felt connection to nature.

My examination of the social experience of animals has aimed to go beyond a demonstration that animals are "socially constructed." Though our relationships with animals do indeed *reflect* the social world in many ways, I have argued that they also *shape* the social world. The former position is easily integrated into the traditional sociological paradigm, but the latter poses a bit of a challenge. The field of sociology seems to tacitly endorse the Nature Lost premise and use it as a justification for ignoring the natural world. It is only in primitive societies, the thinking goes, that nonhumans played a critical part in organizing people's sense of self, their social relations, and the structure and culture of society. Technology for the most part liberated "modern" society from nature. Though the environment may be recognized as a key input that fuels society, most social scientists seem to think that the primary aspects of our social world can be understood and explained without reference to the biophysical environment.[24]

Through the humble pigeon, I have endeavored to depict the *dialectical* relationship between the human and nonhuman world, particularly at the level of what Erving Goffman called "the interaction order." By this term, he meant to draw attention to the question of how individuals coordinate interaction and make meaning in socially situated face-to-face encounters. Following George Herbert Mead, Goffman believed (as do most scholars of interaction today) that the "orderliness" of interaction "is predicated on a large base of shared cognitive presuppositions."[25] By making shared understanding (expressed through language) the sine qua non of interaction, this perspective presumes that animals are not a part of the interaction order. I challenged this axiom in my analysis of pigeon feeding in Father Demo Square, where pigeons and people coordinated behavior without intersubjectivity and, over time, even developed their own unique interactive routines and "role" expectations. In showing how encounters with pigeons influenced people's behavior in, and interpretation of, public space in Father Demo Square and Piazza San Marco, I argued that pigeons became a constitutive feature of the interaction order of these spaces. Even in Trafalgar Square, pigeons greatly contributed to people's perception of, and relation to, the space and each other. Coding pigeons as taboo, hiring park wardens to enforce a feeding ban, and having a falconer patrol the square all served to inscribe Mayor Livingstone's definition of the square as an orderly "cultural space."

Mead and Goffman rightly stated that one's sense of self is created through interaction—ergo the concept of the "social self." Following this logic, I showed how even interactions with nonhumans can play a part in organizing our sense of self and our social relations. The New York rooftop flyers made distinctions among themselves that were grounded in the "natural" differences of their birds, strongly identifying as "flight men" or "tiplet men" and gaining or losing status in the group based on others' perception of their birds' quality. So strong was their identification with their pigeons that these men forged primary ties that transcended the racial, ethnic, and geographic divisions that remained salient in other spheres of their social lives. In Berlin, the social situation of the homesick Turkish immigrants led them to use pigeons to define and perform their ethnic identity in novel ways and to forge primordial social ties. I drew parallels between these groups' animal practices and totemism to stress the fact that it is not only in "primitive" societies that nonhumans can organize people's identity and social world.

Clifford Geertz described the cockfight in Bali as a dramatized "simulation of the social matrix," a story the Balinese told themselves about how their society is organized. In looking at pigeon racing in the Bronx and

South Africa, I revealed that animal competition remains a compelling site of social dramatization in "modern" societies. Furthermore, I contended that pigeon racing did not simply mirror the social matrix—pigeon racing influenced the stories that fanciers told about themselves and their social world. From their rooftops, fanciers in the Bronx were able to playfully pantomime the mythic struggle between man and nature, immersing themselves in a magical world where they were the playthings of hostile and untamed elements. In winning back their birds, they gained a sense of control over their environment. In South Africa, the Million Dollar Pigeon Race explicitly modeled meritocratic and neoliberal ideals of equal opportunity and free competition, a story that resonated deeply, given the setting in postapartheid South Africa and the fact that it brought together competitors from around the globe. Moreover, I found that the MDPR actually played an important part in convincing some fanciers that our globalizing society is in fact reaching these ideals. Nonwhite South African fanciers, for instance, held up their own participation in the MDPR as proof that their nation was living up to its promise to be color-blind. And the fact that a commoner could go head to head with the Queen of England signaled the decline of class and status barriers in the global era. Through the pigeon, fanciers participated in the construction of the heralded "flat world." But the MDPR also is a story about how rationalization seems to be disenchanting "the pigeon game" and how "free" competition has exacerbated inequality: a division of labor estranges fanciers from their birds, money trumps skill, and the poor are priced out.

Hybridity as a Way of Life

In the end, we are left with a central paradox of urbanized society's relationship to the environment: we believe that encounters with "nature" transcend social life, yet our experience of nature is profoundly social.[26] *Seeing the social in nature* appears threatening to many environmentalists and nature lovers because it breaches the barrier between sacred and profane worlds. In doing so, it seems to undermine one of the primary warrants for conserving the environment. Indeed, the asocial-nature ideal has played an important role, especially in the United States, in convincing governments to protect endangered species and set aside "pristine" tracts of land as preserves. However, as William Cronon has noted, by reproducing "the dualism that sets humanity and nature at opposite poles," this quixotic standard does not offer a blueprint of "what an ethical, sustainable, *honorable* human place in nature might look like."²⁷ As James Gibson acknowledges, notions of purity can foster apathy for the "hybrid" landscapes most of us actually live in. As intimated above, notions of the "wild" as untainted by society can also impede appreciation for the "hybrid" animals (i.e., synanthropes) that inhabit our hybrid landscapes.

Seeing the social in nature need not threaten our desire to associate with and protect nonhuman life. For instance, the Turkish immigrants' sense of primordial attachment to their pigeons, and to nature in general, resulted from locating *nature* within the social category of *nation*. And in Piazza San Marco, it was precisely because the street pigeons there had been designated as cultural objects that visitors, many of whom claimed to dislike pigeons, delighted in feeding them. Interacting with pigeons connected them to the urban milieu, rendering the space a personally meaningful place. Even those who see relations with animals as a means of communing with nature, such as bird-watchers, demonstrate the interdependence of asocial and social nature experiences. For many birders, part of the excitement and meaning of spotting a rare bird comes from recording the observation and sharing the encounter with others who have a similar interest. I witnessed a similar phenomenon when the hawks Bobby and Violet nested on NYU's library. A community of sorts sprang up around the hawks in the form of an online chat room and in-person "meetups" in Washington Square. People were eager to be the first to report a piece of news, such as the baby hawk leaving the nest, and a few of the "regulars" gained status from other birders for their keen observational skills. Conversely, a few of the meetup members told me that they had never really paid much attention to urban wildlife before. For them, the social gatherings at the park and the group explorations of the nearby environs played a part in abetting their interest in the hawks and the surrounding ecology.

The journalist Robert Sullivan writes that nature is prospering in cities like New York. The five boroughs contain a greater variety of habitats than the surrounding suburban and rural areas, with their uniform tracts of grass and fields, and Jamaica Bay (Queens) contains more species of birds than Yellowstone and Yosemite Parks combined. Most of us have not noticed this nature, he argues, because it is not the kind we are looking for—it is less precious, less "pure."²⁸ Who would bother to detect the biodiversity lurking in interstitial spaces such as highway embankments, particularly if they are strewn with litter? These species, many of which have learned to adapt their behaviors to the habits of people, powerfully demonstrate the extent to which the social and the natural are literally—not just conceptually—intertwined.

Given the pace and scope of human-induced ecological disruptions, as well as efforts to save or restore open spaces in and around urban regions, ecologists predict a future where more and more species will fashion their survival strategies around humans. Coyotes and other predatory animals, which most of us think "belong" in the country, now regularly encroach the city limits and find our streets to be suitable habitats. Whether or not humans feel that we have any ethical responsibility for the dependent masses of urban animals that we haphazardly abet, we should recognize that their presence does not reflect the incomplete removal of nature from the city but rather the progressive adaptation of nonhumans to an artificial environment. Poison will not stop them. While the idea of the city as a setting devoid of human encounters with nonhumans was always a mirage, the rise of synanthropy, the "greening" of cities, and the blurring of urban and rural boundaries as a result of sprawl shatters the hackneyed "imaginative geography" of the city. If "urbanism as a way of life" described a peculiar set of interactions and attitudes that made the social order of the city distinct from the country, then perhaps we should now speak of "hybridity as a way of life" to denote the fact that nonhuman species are regular participants in what Jane Jacobs called the "intricate sidewalk ballet" of the urban social order.

Learning to appreciate the "contaminated" biodiversity of our hybrid landscapes, an attitude which might be better fostered by seeing the social in nature rather than holding species up to an asocial template, can expand the terrain of environmental conservation and may even offer clues for how to model more sustainable human and nonhuman cohabitation. Perhaps such appreciation could even begin with that most pedestrian creature of all, the pigeon.

ACKNOWLEDGMENTS

THIS BOOK GREW out of an ethnographic methods class that I took at the City University of New York in my second year of graduate school. I will always remember that fateful day when I read aloud to the class a rather mundane excerpt of my field notes about people who fed pigeons in Father Demo Square. Mitchell Duneier, the instructor, approached me afterward and said, 'I'd like to take you to lunch to talk about the pigeons.' From that moment forward, Mitch's heroic efforts guided this research and my professionalization into the discipline. He believed in this project when few did, including myself. A prototypical mentor and ethnographer, he set an example that I strive to emulate in my own research and teaching. Bill Kornblum also supported me with gusto. His love and immense knowledge of all things related to urban communities was an invaluable resource. Julia Wrigley played a leading role in shaping this work as well. Her keen intellect, academic openness and curiosity, and fascination with birds brought a fresh perspective. In addition to my advisers, there was a constellation of phenomenal professors and fellow students at the CUNY Graduate Center that gladly advised, encouraged, and critiqued me, especially Paul Attewell, Phil Kasinitz, David Goode, Patrick Inglis, and Robert Turner. The Graduate Center is a special place, and I look back on my time there with a sense of nostalgia matched only by the pride I feel for being affiliated with this great public institution.

I continue to be humbled by the extraordinary scholars who showed an interest in me when I still felt as though I didn't know what I was doing. A special thanks goes to Jack Katz, who over the years has gently disciplined me to be a more careful and thoughtful sociologist, and who encouraged the University of Chicago Press to take a chance on my book when it was only half-baked. Howie Becker masterfully played the roles of the sage and the curmudgeon, assuring me in my moments of confusion that I already had it all figured out and tenderly chastising me when my ethnographic accounts became clouded by my theoretical pretensions. I remain indebted to Elijah Anderson, whose ethnography conferences were critical to the development of my research, and whose encouragement over the years has always buoyed my spirits.

The task of turning my thesis into a book began at Harvard University, where I was fortunate enough to be housed for a year and a half as a Robert Wood Johnson Foundation Scholar in Health Policy. I owe much of the book's development during my time in Cambridge to fellow "young ethnographers" who read and critiqued a draft of the manuscript: Claudio Benzecry, Ryan Centner, Andrew Deener, and Jonathan Wynn (Book Club 1). I also benefited from stimulating discussions with Patricia Strach, Michèle Lamont, and Chris Winship.

When I moved to NYU, I (unfortunately) still had a long way to go with the book. But I could not have found a more felicitous milieu to complete the task. My joint appointment in environmental studies and sociology has produced a wonderful cross-fertilization that, I hope, is apparent in these pages. In both homes I have found incredibly supportive, insightful, and patient colleagues who, over many an espresso and hallway conversation, have helped my ideas and writing mature. And the courses I have taught on the environment, human-animal relations, and ethnography have also profoundly shaped this manuscript. A transformative moment for my book came when, thanks to the generosity of Jeff Manza and the NYU Sociology Department, I convened an "author meets critic" workshop with David Grazian, Ruth Horowitz, Eric Klinenberg, Harvey Molotch, Rob Smith, and Sudhir Venkatesh. Harvey, who supported this project long before he was my colleague, graciously moderated as these folks spent an afternoon deconstructing my manuscript. Though I walked away from the meeting a bit overwhelmed, I returned to the book with renewed vigor and clarity. It was because of their efforts that I finally turned the corner and tamed this beastly book. In my time at NYU, I have once again benefited from kindred "young ethnographers" who read and commented on a full draft: Shamus Khan, Jooyoung Lee, Erin O'Connor, Harel Shapira, Tyson Smith, Iddo Tavory, and Lucia Trimbur (Book Club 2). Other good friends who have read all or parts of the manuscript and provided invaluable written and verbal comments are Matt Desmond, Alice Goffman, and Ed Walker.

At the University of Chicago Press, I will forever treasure Doug Mitchell's unending exuberance. More than an editor, the man is an institution. It has been an unmatched privilege to develop the book under him and, of course, to commune in person and over e-mail about anything having to do with pigeons and drummers named Doug. Tim McGovern, the puppeteer behind the scenes, has also been a treat to work with. If anything, the fact that this book has finally been published means that my ritual visits with Tim at the ASA book fair will no longer be tinged with guilt. I am also grateful to Joann Hoy, whose eagle eyes caught the many remaining grammar errors lurking in the penultimate version of the manuscript.

Ethnography is possible only because people let us into their lives. There is no way for me to repay that generosity, for this enterprise has likely enriched me far more than my participants. Though I do not expect those I have written about to agree with all of my interpretations, I hope that they at least know that I made an earnest effort to see the world through their eyes. Several of the "pigeon guys" in New York deserve special mention. From the day I wandered into Broadway Pigeons and Pet Supplies in Brooklyn with the hopes of learning about rooftop pigeon flyers, Joey Scott welcomed me and brokered access to the rest of the group. Carmine Gangone was the first to invite me onto his roof, and I cherish the countless afternoons that I spent up there with him and Frankie. Though Carmine is no longer able to terrorize the skies, I hope that he finds this book—including the magnificent pictures taken by Marcin Szczepanski—to be a worthy record of his craft. In the Bronx, Marty McGuinness excitedly brought me into the world of homing pigeon racing and treated a "kid" with dreadlocks and no knowledge of "the pigeon game" with the utmost respect. Beyond New York, my research relied even more heavily on the generosity of others. In Berlin, Beyhan Yildirim selflessly volunteered to serve as my translator, guide, and host for over three weeks. Ahmet Çabakçor enthusiastically took on the role of my primary informant, and Oliver Hartung took some evocative pictures. In South Africa, Theo Peters and his wife, Hillary, volunteered to house and feed someone they had never met for over three weeks, after which Wim Peters happily took me to Sun City and allowed me to stay in his hotel suite. The Peters brothers were also crucial in my recruitment of participants. Though I do not have the space to thank all of the people who agreed to be a part of this project, I hope that they sense my gratitude when they come across their names on the printed page.

I close by emphatically thanking all of my friends and family. Doug Porpora patiently mentored me when I was a precocious undergraduate and almost single-handedly steered me toward my destiny. Since that time, he has become a dear friend and colleague. Though I lived in a virtual cave for six years in Brooklyn while completing graduate school, my roommates at #425 and friends at the "Chicken Hut" made sure I had some fun along the way. My lovely wife, Shatima, has more than tolerated this project for our entire relationship. I had to cut our first date short to go to a pigeon race; I missed her graduation because I was at the Million Dollar Pigeon Race; and over the subsequent years, some of our dates have included trips to Joey's pet shop and Carmine's rooftop. Beyond emotional support, she has provided smart and critical commentary over many a dinner. My mom and dad did everything in their power to give me the best possible life chances, and their pride in my accomplishments warms me. My brothers, Doug and Ian, are my best friends, and our bond protects me. Even in death, my grandmother Blanche—aka Nanie—continues to be my special buddy. And, finally, may my beloved grandparents Anna and Hugo rest in peace and look with satisfaction upon the successes of their descendents that they enabled.

NOTES

Introduction

1. Douglas 1966.
2. This moniker is a remnant from the days when people kept pigeons in their attic lofts.
3. Wirth 1938, 1. For a critique, see Latour 1993; Philo and Wilbert 2000.
4. Gerald Suttles (1968) describes how ethnic groups in Chicago viewed their neighborhoods as "defended territories" in which they framed the encroachment of "outsiders" as an "invasion."
5. In the end, "Parker the Pigeon" lost out to "Pearl the Squirrel."
6. Haag-Wackernagel and Moch 2004, 307.
7. See Povoledo 2008; Brooklyn councilman Simcha Felder's 2007 report, "Curbing the Pigeon Conundrum." Blechman (2006, 153) reports that the US Department of Agriculture kills 75,000 pigeons per year.
8. Pigeons can produce over a half-dozen "clutches" of two eggs each year, generally breeding up to the food supply (see Johnston and Janiga 1995). Regarding symbolism, "There is good evidence that much of the iconic imagery of doves we still recognize today was originally based on the same species as our pesky street pigeons" (Humphries 2008, 3); see also Blechman 2006.
9. See Soeffner 1997, 99. For more on pigeon history, see Levi 1941.
10. Secord 1981, 170.
11. Secord 1981, 165; Nicholls 2009, 790; Darwin (1859) 1909.
12. See Marzluff and Angell 2005, who discuss another paradigmatic synanthrope—crows.
13. Berger 1980, 21.
14. Wilson 1993, 32.

15. Louv 2005; Pyle 1993, 147.

16. Pyle 1993, 145. See also Abram 1997; Kellert and Wilson 1993.

17. For a review, and rebuttal, of "Community Lost," see Wellman 1979.

18. Gibson 2009, 8; see also Colomy and Granfield 2010.

19. Gibson 2009, 9; Weber (1918) 1946; see also Wolch 1998.

20. M. Bell 1994, 138.

21. Cronon 1995, 25. On "asocial nature," see Angelo and Jerolmack 2012.

22. M. Bell 1994, 138.

23. See Arluke and Sanders 1996; Fine 1997; Greider and Garkovich 1994; Irvine 2004.

24. M. Bell (1994, 4) calls his study an exploration of "the social experience of nature."

25. Lévi-Strauss 1963, 89–90; Durkheim (1912) 1995.

26. Latour 1993. Environmental sociologists William Catton and Riley Dunlap (1980) refer to the assumption that society is liberated from nature as the "Human Exemptionalism Paradigm."

27. Marzluff and Angell 2005, 7.

28. See Angelo forthcoming; Kosut and Moore forthcoming; Sullivan 2010; Tissot 2011.

29. See Arluke and Sanders 1996; Fine and Christoforides 1991; Herda-Rapp and Goedeke 2005.

30. See Blumer 1969; Mead (1934) 1967. For critiques, see M. Bell 1994; Sanders 2003; Irvine 2004.

31. Geertz 1973.

32. Philo 1998, 52. See also Čapek 2010.

33. Tuan 1984, 3.

34. See Anderson 2011; Cressey (1932) 2008; Duneier 1999; Jacobs 1961; Kornblum 1974; Park and Burgess 1925; Rieder 1985; Simmel 1971; Suttles 1968; Wirth 1938; Zorbaugh 1929. Erving Goffman's *Relations in Public* (1971) masterfully deconstructs the tacit rules that enable urbanites to go about their lives among a sea of strangers with some surface agreement of trust and a large degree of order.

35. Tsing 2005, xi.

36. Goffman 1983.

37. Friedman 2005.

38. See Arluke and Sanders 1996; Irvine 2004.

39. Marcus 1995, 106; Appadurai 1986, 5.

40. Seligmann 2004, 11; on the humanistic and scientific merits of disclosure, see Duneier 1999.

Chapter 1

1. Jacobs 1961. Father Demo Square also abuts the blocks where Mitchell Duneier (1999) documented the lives of unhoused book and magazine vendors.
2. Belguermi et al. 2011; see also Weber, Haag, and Durrer 1994.
3. Jacobs 1961, 50.
4. On sociability, see Simmel 1949; on the "interaction order," see Goffman 1983. For those readers versed in "actor network theory," we can think of pigeons as nonhuman "actants" that, like humans, contribute to the orderliness found in this space (see Latour 1988).
5. Goffman 1963, 24.
6. Goffman 1963, 126; Robins, Sanders, and Cahill 1991, 3. Duneier (1999) shows how some of the black street vendors he studied took advantage of this possibility to "entangle" passing white women in conversation that they sought to avoid.
7. On sustaining flow and "play membranes," see Csikszentmihalyi 1975; Goffman 1961.
8. Goffman (1974) 1986, 43.
9. For a more theoretical answer to this question, see Jerolmack 2009a.
10. See Blumer 1969; Mead (1934) 1967.
11. Irvine 2004; Sanders 1999; the original statement of this problem comes from Bateson (1956).
12. Mitchell and Thompson 1986, 1990.
13. See Goode 2006; Mechling 1989. The same is true for people; see Fuchs 1989; Garfinkel 1967; Goode 1994; Hendriks 1998.
14. On the role of inefficient gestures in play, see Miller 1973.
15. Watson and Potter 1962; see also Riesman and Watson 1964.
16. Buber 1970; Myers 1998, 82; see also Owens 2007.
17. Simmel 1949.
18. See Anderson 2011; Duneier 1999, 1994; Jacobs 1961; W. H. Whyte 1988.
19. Goffman 1963, 132. Jack Katz (1999, 326) calls such contexts as these, in which he includes places as varied as supermarket checkout lines and highways, areas of "cultural communism." In such locales, "no externally relevant status is routinely useful in the distribution of rights."
20. In his study of relations between unhoused vendors and wealthy neighborhood residents, Duneier (1999) focuses on this tension on the very streets where my fieldwork was carried out.
21. Weber, Haag, and Durrer 1994, 58; see also Humphries 2008, 153–72.
22. Kellert and Wilson 1993; Gibson 2009; Pyle 1993. I examine "biophilia" in chapter 3.

23. Goffman 1967, 162.

24. Jacobs 1961, 56; on the "blasé attitude," see Simmel 1971.

Chapter 2

1. Katz 1997, 412–13.

2. The term "falconer" is employed even if the bird used is a hawk, as it is in London.

3. M. Bell 1997, 824; see also Gieryn 2000.

4. Littlewood 1991, 117; James Morris 1974, 81; see also Madden 1964.

5. Littlewood 1991, 117–18.

6. James Morris 1974, 82.

7. James Morris 1974, 81; Jan Morris 1982, 27.

8. Jan Morris 1982, 27–28.

9. See Kerr 1985; Gumbe 1995; Willey 1995.

10. *Birmingham Post* 1999; see also O'Reilly 1999.

11. C. Bell 1991.

12. On the ways that tourists' expectations of place are constructed from afar, see Urry 1990.

13. The English-language sign cited "Article 23 of the regulations of the Metropolitan Police."

14. Urry 1990.

15. See Gieryn 2000; Latour 1988.

16. Davis and Marvin 2004, 76.

17. Wynn 2011.

18. On the search for authenticity in entertainment and tourism, see Grazian 2003; Urry 1990.

19. On "imaginative geographies" of animals, see Philo and Wilbert 2000; Wolch and Emel 1998.

20. Barry 2006.

21. This is a pseudonym.

22. http://europeforvisitors.com/venice/articles/pigeons_of_venice2.htm.

23. The popular tourist website Veniceword.com also dryly referred to the pigeon paradox as "the Italian way of solving problems" (2007).

24. A Venetian Office of Commerce worker told me that people with disabilities were granted licenses to sell feed and souvenirs like Murano glass over 40 years ago. The city ceased granting licenses after being overrun with vendors, but those licenses already granted had 'old authorization that we can't cut off.' She lamented that an unintended consequence of restricting licenses is that those who own them are 'very, very rich.' See Barry 2006; Moore 2007.

25. Mace 2005, 12.

26. I never was able to find any official documentation that might reveal the origin of the feed-vending licenses. I combed through the Municipal, House of Lords, and Westminster archives to no avail; librarians and employees sent me in circles.

27. McCullin 2005, foreword.

28. Raphael 1967.

29. Johnson 1986.

30. Barwick 1996.

31. *Evening Standard* 1997b.

32. *Evening Standard* 1997a.

33. McCarthy 2000.

34. The borough of Westminster, which encircles the square, has city status.

35. *West End Extra* 2000; Mason 2000.

36. Nettleton 2000.

37. Livingstone 2001.

38. Jenkins 2001.

39. The license was under the jurisdiction of the Department of Culture, Media, and Sport.

40. Alleyne 2001.

41. Some details of these events, cross-checked with news stories, come from an interview with Neil Hanson, the public relations person for Save the Trafalgar Square Pigeons.

42. A sign in the square, "Acts Prohibited within the Squares," states under bylaw 3A: "No person other than a person acting on the direction of the mayor shall within Trafalgar Square (1) feed any bird (which shall include dropping or casting feeding stuff for birds); or (2) distribute any feeding stuff for birds."

43. Blechman 2006, 141.

44. Lydall 2006a, 2006b; Muir 2006.

45. Because I sought to remain a neutral observer in the eyes of the heritage wardens and sanitation workers, I assisted in feeding only when they appeared to be preoccupied.

46. A DVD called *Share the Blue Skies* features scenes from Trafalgar Square before the feeding ban, including images of this man directing feeders in front of Rayner's feed stand.

47. *Daily Telegraph* 1965.

48. All of my efforts to speak to *any* official GLA representative, including months of e-mails and phone calls and three personal visits to City Hall, were rebuffed.

49. Stewart 2008.

50. Taylor 2007.

51. M. Bell 1997, 830.

52. *New York Times* 1966, 49. After searching many news media and popular culture outlets, this is the oldest reference I have found for pigeons as rats with wings. The term went Broadway after Woody Allen used it in his 1980 film *Stardust Memories* (see Jerolmack 2008).

53. Philo and Wilbert 2000, 22.

54. Philo and Wilbert 2000, 11; Philo 1998, 52; see also Sabloff 2001.

55. See Angelo and Jerolmack 2012; Cronon 1996.

56. Douglas 1966, 12.

57. I analyze this discourse in relation to pigeons in Jerolmack 2008. On New York's paradigmatic efforts to clean up "disorder" and improve the quality of life, see Vitale 2009.

58. *Hackney Gazette* 2006; Sewell 2007.

59. HC/CL/JO/10/645/368.

60. On the homogenization of urban space and the enchantment of place, see Wynn 2011.

Chapter 3

1. Mooney 2008. While in 1990 "whites outnumbered Hispanics in the area two to one, and the Asian population was less than 10%" (*Queens Tribune* 2006), by 2009 Hispanics outnumbered whites two to one and the Asian population was 24%; half the residents were foreign born, and the median household income was a respectable $42,502 (http://www.census.gov/acs/).

2. Tyson, who grew up in a Brooklyn ghetto, says he realized his pugilistic prowess at age ten when he pummeled an older kid who killed his pigeon. Upon getting demolished in his final professional bout, Tyson famously declared that he wanted to return to his pigeons.

3. I cut off my long hair in July 2005. Both before and after, my dreadlocks were a source of curiosity, conversation, and sometimes derision.

4. Levi 1941, 41–42; see also Bodio 1990; Kligerman 1978; Levi (1965) 1996.

5. Brooklyn Community District 18 profile, from http://www.nyc.gov.

6. In 2000, this relatively racially stable neighborhood was 49% black and 38% Hispanic. Over three-quarters of residents were renters. The median household income was $31,249 (the median for New York City was $38,293). See http://www.nyc.gov.

7. On "la guerra" in Brooklyn, see Kligerman 1978; Schwartz 1986.

8. The flyers often referred to each other by occupation, even long after one had retired.

9. Such distinctions illustrate the real-world messiness of social categories like class. Weber ([1918] 1946, 180) highlighted status and "styles of life" to combat oversimplified notions of class as market position. I do not intend to reify class; rather, I am

foreshadowing a performative notion of class to be fleshed out below (see Grasmuck 2005; Sherman 2007).

10. On working-class habitus, see Bourdieu 1977; Desmond 2007; Willis 1981.

11. On dirty work and stigma, see Hughes 1958; cf. Ashforth and Kreiner 1999.

12. Matthew Desmond (2007, 44) makes a similar argument about the ways that the working-class rural backgrounds of forest firefighters conditioned their bodies and minds for their work.

13. Harper 1987, 167. See Kefalas 2003; Kornblum 1974; Lamont 2000. Chapter 5 revisits this theme.

14. This is a pseudonym. Carlos requested anonymity because he preferred not to let his professional colleagues know that he kept pigeons.

15. Kefalas 2003, 5. Kefalas emphasizes how clean homes, cultivated gardens, and well-maintained properties are a means for the working class to express their moral values.

16. Soeffner 1997, 108.

17. For more on this distinction, see Liebow 1967; Bourgois 1995.

18. Kellert 1993a, 21.

19. Wilson 1993, 31.

20. Kellert 1993b, 43; 1993a, 11. On premodern biophilia, see Nelson 1993.

21. Gibson 2009, 9, 11.

22. Ibid., 3.

23. Irvine 2004, 32.

24. Kellert 1993b, 42.

25. Tuan 1984, 25.

26. Bodio 1990, 47.

27. Katz 1999, 33.

28. Nash 1989, emphasis added.

29. Mead (1934) 1967, 175, 154.

30. Irvine 2004, 162–66; Sanders 2003, 412. M. Bell (1994) studies how the social self is shaped through nature.

31. On nature as a reflection of social life, see Greider and Garkovich 1994; on how nonhumans organize the social realm, see Latour 1993.

Chapter 4

1. Fine 2003, 21.

2. The men did not make ethnic claims that might be at odds with national claims (e.g., identifying as Kurdish). Thus, in this case's analysis of ethnicity, nationality is implied.

3. Gibson 2009, 11; cf. M. Bell 1994; Fine 2003.

4. Turks in Germany, though, are not as socially or economically marginalized as similar migrant groups in neighboring France (see Böcker 2004; Ehrkamp 2005). Many eligible Turks refuse to give up Turkish citizenship in exchange for German citizenship. Yet this aversion to German nationality may be a partial by-product of historical laws. Until 2000, citizenship in Germany was determined by blood, not by territory; generations of German-born immigrants did not have an open path to citizenship (see Brubaker 1992; Caldwell 2007).

5. Some estimates place the number of unemployed Turks at 20–40%, though the numbers are unreliable (see Böcker 2004; Dave, Ivanova, and Sutton 2006; Hönekopp 2003; Mueller 2006).

6. The exceptions were three men who were actually born in Germany, but who spent most of their childhoods in Turkey, and four men who came as young adults in search of better employment.

7. See Hillman and Rudolph 1997.

8. In the Middle East and Europe, tumbling competitions were once common (Levi [1965] 1996).

9. Severe restrictions on importing birds into the European Union, especially after the outbreak of the avian flu, meant that the birds brought in had to be smuggled. See European Commission Decision 2008/592/EC, http://eur-lex.europa.eu/LexUriServ/LexUriServ.do?uri=CELEX:32008D0592:en:NOT.

10. Geertz 1963, 109.

11. Beeley 1970, 476; Ehrkamp 2005, 354.

12. Several of the caretakers *did* smoke, and I did not see others discourage them.

13. Tragically, Hasan died of cancer in 2010.

14. Cronon 1996, 16; Gibson 2009, 28, 153; M. Bell 1994, 7.

15. Suttles 1968, 16. On boundary work, see Lamont 2000.

16. Gibson 2009, 39, 11–12; see also Kellert and Wilson 1993.

17. Lévi-Strauss 1963, 89; Durkheim (1912) 1995, 284.

18. The New York pigeon flyers also valued bloodlines and formed an attachment to the entire stock, but these bloodlines are were not conceived of in ethnic or national ways.

19. Geertz 1963, 259; see also Shils 1957; Schmalenbach (1927) 1977.

20. M. Bell 1994, 87–88; see also Cressey (1932) 2008; Park and Burgess 1925; Wirth 1938.

21. On urban villages, see Gans (1962) 1982; Suttles 1968. On the rural-urban imaginary, see M. Bell 1994; Williams 1973. For a critique of classic community paradigms, see Viddich and Bensman 1958; Wellman 1979.

22. For an extended version of this argument, see Jerolmack 2007. See also Brubaker et al. 2006; Gil-White 1999, 2001; Grosby 1994.

23. There *is* a history, however, of breeding homing pigeons for racing, and a variety of fancy pigeons for competitive exhibitions, in Germany. As elsewhere, it is a hobby in decline.

24. Of the caretakers' two-dozen friends who hung out at the coops but did not have birds, most said they had tumblers in Turkey and aspired to have them again but did not have the resources.

25. Waters 1990.

26. See Lassiter and Wolch 2005.

Chapter 5

1. According to the US census, Bushwick—once a prominent Italian neighborhood—went from 100% white in 1950 to 3% white, 24% black, and 67% Hispanic in 2000. It, along with "Bed-Stuy," was devastated by riots during an electrical blackout in 1977. In 2000, Bed-Stuy—an iconic African American neighborhood and the setting for Spike Lee's film *Do the Right Thing*—was 1% white, 18% Hispanic, and 77% black. This contiguous area is experiencing gentrification on the fringes that touch Fort Greene, to the west, and Williamsburg, to the north.

2. The conflict that can result from racial succession is poignantly documented in Jonathan Rieder's *Canarsie* (1985). For more on parochial communities, neighborhood transition, and racial conflict, see Kefalas 2003; Pinderhughes 1997; Suttles 1968; Wilson and Taub 2007.

3. Cooley 1909, 23; see also Anderson 1978. The term "we-feeling" also comes from Cooley.

4. Anderson 1978, 191.

5. Brown-Saracino 2011, 361; on "master status," see Hughes 1945. Anderson (2011, 189–215) usefully unpacks "ethnocentric" and "cosmopolitan" orientations; see also Jerolmack 2009b.

6. Anderson (1978) describes how a primary group formed by poor black patrons in a bar afforded members a chance to "be somebody."

7. Anderson 1978, 37, 17. In *A Place on the Corner*, Anderson examines how working-class black men who gathered at a tavern competed for status in the group through sociability.

8. Goffman 1959, 12; Katz 1999, 351.

9. Katz 1999, 72.

10. Harper 1987, 195. In the 2008 movie *Gran Torino*, Clint Eastwood's character stereotypically demonstrates this working-class male custom in his interactions with his barber. For more on this ritual, see Anderson 1978; Bauman 1972; Lee 2009.

11. Calling on "witnesses," including me, to verify assertions was a common tech-

nique. The men seemed to assume that I was an impartial "outsider" and that I was a competent judge of pigeons.

12. Play commonly reverses the meaning of "serious" behaviors (Goffman 1961; Lee 2009).

13. Goffman 1959, 75.

14. Duneier (1999, 49–52) makes a similar point about how unhoused street vendors invoke a "fuck it" mentality to frame their social situation as a choice rather than an inevitability.

15. Similarly, Grasmuck (2005, 134–38) shows how little league baseball coaches risked scorn if they displayed hypermasculinity without displaying knowledge of the game.

16. See Zerubavel 1997, 32–33.

17. Lévi-Strauss 1963, 101.

18. Anderson 1978, 209.

19. Park 1950, 250.

20. Goffman 1959, 35; Myerhoff 1980, 150.

21. Almost all of the old-timers and blacks used "Spanish," not "Hispanic" or "Latino," as a racial label. Some Hispanic men used this descriptor as well. This usage has a long history in New York.

22. See Anderson 1990; Rieder 1985; Suttles 1968.

23. Kornblum 1974, 9, 26.

24. See Brown-Saracino 2011.

25. As I was in the final stages of writing up this book, I discovered an essay written by a German sociologist that described a similar process in a very different time and place. Based on his World War II–era childhood recollections of a community of miners who raced homers in the Ruhr district, Hans-Georg Soeffner (1997, 113) argues that pigeons were an appropriate "flag of the clan" because they hovered above the religious, ethnic, educational, and territorial differences that threatened to cause tension among this gritty group of migrant workers. Relatedly, M. Bell (1994) found that class tensions threatened to undermine social solidarity in the village he studied, and that grounding their sense of selves in nature was a way for villagers to transcend these social divisions and rally around an identity that they could all share.

26. Sal and Carlos, the only white-collar flyers, were also the only flyers to comment on diversity. Both said that one benefit of keeping pigeons is that you meet "all kinds of people."

27. Hunter, 1974, xiv.

28. Grasmuck 2005. Different ethnic and age cohorts sometimes brought distinct flavors to the masculinity script; for example, some of the younger men of color added flourishes, in certain contexts, that scholars might label as "street" (Anderson 1999) or "machismo" (Bourgois 1995). But underneath this tip was the iceberg of a more general, shared working-class masculinity. Like Gans ([1962] 1982, 277–78), I "continue to be

impressed by how similarly members of different ethnic groups think and act when they are of the same socio-economic level . . . and must deal with the same conditions."

29. See Liebow 1967.

30. Kornblum 1974, 75; see also Anderson 1978; Duneier 1994; Liebow 1967; Suttles 1968; W. F. Whyte (1943) 1993; Young and Willmott (1957) 1992.

31. Kornblum 1974, 69; see also Anderson 2004, 2011.

Chapter 6

1. Geertz 1973, 436, 440. On other types of "social dramas," see V. Turner 1974.

2. Durkheim (1912) 1995; studies of gambling (Avery 2009) and stealing (Katz 1988) suggest that the charm of these forms of "action" centers on sensual thrills, not remuneration (Goffman 1967).

3. See http://www.ifpigeon.com/; Yakin 2008. No good data are available on the number of fanciers worldwide.

4. Kefalas 2003. Some of the fanciers emigrated from Italy, eastern Europe, and the Caribbean.

5. See Biro et al. 2006; Blechman 2006, 8; Humphries 2008, 74–77.

6. Goffman 1967, 179.

7. See Bodio (1990, 26–27), who describes basketing in Boston during the 1950s.

8. On liminality, see V. Turner 1974.

9. Some fanciers in other clubs still used manual clocks. In that case, their birds wore a rubber leg band (countermark) rather than an electronic chip. When the bird came home, the fancier had to trap it, remove the countermark, and place it in the tamper-proof clock that then stamped the arrival time on a piece of paper. When clocks reached the club, race officials opened them and manually logged each time. Electronic clocks enabled quicker and easier logging of arriving birds, and they also protected against cheats. Pigeon lore was filled with instances of fanciers circumventing the protective seals of their clocks in order to stamp an earlier arrival time.

10. Though few of the fanciers spoke Spanish, this homophobic slur was a common insult.

11. Harper 1987, 195.

12. On one occasion, Bratton, who retired from the NYPD in his 40s with a full pension, was puzzled over my own career choice. Others chimed in, explaining that if I had become a fireman at the time I entered graduate school, I could be making "good money" by then. They added that, with my education, I could work my way up to chief and retire early.

13. Pawson 1996; Peters 2000. New research shows that social hierarchies spontaneously emerge within flocks, with dominant birds dictating the group's flight path (Nagy et al. 2010).

14. Geertz 1973, 440.

15. Genetic tests suggest that wayward homers have "made substantial contributions to feral pigeon populations" (Stringham et al. 2012, 1).

16. One fancier writes that a good pigeon should "feel like a wedge between your hands, with broad shoulders and full breasts tapering toward a slim tail section" (Bodio 1990, 34).

17. See http://www.cjccombine.com/weather/.

18. Gmelch 2001, 134.

19. Durkheim (1912) 1995; on "we-feeling," see Cooley 1909.

20. Wacquant 2004, 6.

21. Goffman 1967, 260–61.

22. I tossed the chico and corralled birds into Franco's loft on three occasions, experiencing firsthand the cotton mouth, momentary incapacitation, and adrenaline rush.

23. Weber (1918) 1946, 155, emphasis added.

24. Malinowski 1948, 116; see also Gmelch 2001.

25. Rozin and Schiller 1980.

26. In a book widely read by fanciers, the veterinarian Wim Peters (2000, 136) writes that, beyond superb preparation, the winning pigeon possesses a "will to win . . . a determination to carry on until the end—an unwillingness to give in—even to the point of collapse."

27. Katz 1988, 8.

Chapter 7

1. http://www.southafrica-travel.net/north/a1Suncit.htm.

2. Peters 2005, 30. His two books are *Fit to Win* and *Born to Win*.

3. Geertz 1973, 448.

4. Harper 1987, 21; see also Weber (1922) 1978.

5. "Coloured" was an official racial category imposed upon mixed-race people by the apartheid government, regardless of how they self-identified. Coloured people were granted somewhat greater privileges under apartheid than Blacks, such as fewer restrictions on where they could live; today, the majority of people that fall into this racial category speak Afrikaans.

6. Sallaz 2009, xvi.

7. Kinney 2005.

8. Kinney 2005. The facts and figures listed above all come from Kinney's article.

9. To see the 2009 MDPR liberation, go to http://www.youtube.com/watch?v=n658kpk3y9A.

10. See Sallaz 2009, 135.

11. M. Turner 2000.

12. The queen is the honorary head of the Royal Pigeon Racing Association. Blechman (2006, 196) writes, "The Queen's pigeons are descendents of a gift from Belgium's King Leopold to her great grandfather King Edward VII." The queen, though, plays no part in caring for them.

13. Geertz 1973, 435, 449.

14. Friedman 2005.

15. Harper 1987, 21, emphasis in original.

16. Harper 1987, 18; see also Sennett 2008.

17. Peters 2000, 205.

18. Anthony 2003.

Chapter 8

1. http://www.abovebrooklyn.net/.

2. See Richard Lloyd's *Neo-Bohemia* (2005) on the role of postindustrial neighborhoods, and the hip young adults who live and work in them, in the new creative economy. Sharon Zukin's *Loft Living* (1989) shows the role of artists in laying the groundwork for gentrification in New York.

3. http://www.thepigeongame.com.

4. Kligerman 1978, 10–11; Feuer 2001.

5. Kirshenblatt-Gimblett 1983, 199.

6. Robert Putnam (2000) would admire the role that local pigeon clubs played in American society. They once flourished alongside of, and mirrored, other leisure and social organizations that fostered communal and civic ties (e.g., bowling leagues or the Lions Club).

7. This incident is documented in the movie *Up on the Roof*, by J. L. Aronson (2008).

8. Chicago Municipal Code § 7–12-387, § 387(b): It is "unlawful for any person to import, own, keep or otherwise possess any live pigeon within any area designated as a residence."

9. Young and Willmott (1957) 1992, 38.

10. Philo 1998.

11. Philo and Wilbert 2000, 22; see also Douglas 1966; Sabloff 2001.

12. An article written by the world's foremost pigeon biologist concluded that, "in spite of the worldwide distribution of feral pigeons, the close and frequent contact they have with humans, their use as food, and the high prevalence of carriage of human pathogens, zoonotic disease caused by pigeons is infrequent" (Haag-Wackernagel and Moch 2004, 311).

13. For details, see Jerolmack 2008.

14. On chicken coops, see Angelo forthcoming; on beekeeping, see Kosut and Moore forthcoming.

15. See Angelo and Jerolmack 2012.

16. See Wellman 1979.

17. Gibson 2009, 12.

18. Arluke and Sanders 1996, 170.

19. Humphries 2008, 137–38.

20. Tissot 2011; Angelo forthcoming.

21. See Kosut and Moore forthcoming.

22. Appeal from the US District Court for the Northern District of Illinois, Eastern Division, no. 04 C 5429, Harry D. Leinenweber, judge.

23. M. Bell 1994; Cronon 1996, 13.

24. See Catton and Dunlap 1980; Latour 1993.

25. Goffman 1983, 5.

26. This section draws substantially from Angelo and Jerolmack 2012.

27. Cronon 1996, 17.

28. Sullivan 2010.

REFERENCES

Abram, David. 1997. *The Spell of the Sensuous: Perception and Language in a More-than-Human World*. New York: Vintage Books.
Alleyne, Richard. 2001. "Pigeons Lose the Battle of Trafalgar Square." *Times*, February 8.
Anderson, Elijah. 1978. *A Place on the Corner*. Chicago: University of Chicago Press.
———. 1990. *Streetwise: Race, Class, and Change in an Urban Community*. Chicago: University of Chicago Press.
———. 1999. *Code of the Street: Decency, Violence, and the Moral Life of the Inner City*. New York: W. W. Norton & Co.
———. 2004. "The Cosmopolitan Canopy." *Annals of the American Academy of Political and Social Science* 595:14–31.
———. 2011. *The Cosmopolitan Canopy: Race and Civility in Everyday Life*. New York: W. W. Norton & Co.
Angelo, Hillary. Forthcoming. "City Chickens." In *Edges*, edited by C. Calhoun and R. Sennett. New York: NYU Press.
Angelo, Hillary, and Colin Jerolmack. 2012. "Nature's Looking-glass." *Contexts* 11 (1): 24–29.
Anthony, Andrew. 2003. "Don't Fancy Yours Much." *Observer*, May 31.
Appadurai, Arjun. 1986. Introduction to *The Social Life of Things*, edited by A. Appadurai, 3–63. Cambridge: Cambridge University Press.
Arluke, Arnold, and Clinton Sanders. 1996. *Regarding Animals*. Philadelphia: Temple University Press.
Ashforth, Blake E., and Glen E. Kreiner. 1999. "'How Can You Do It?' Dirty Work and the Challenge of Constructing a Positive Identity." *Academy of Management Review* 24 (3): 413–34.
Avery, Jacob. 2009. "Taking Chances: The Experience of Gambling Loss." *Ethnography* 10 (4): 459–74.
Barry, Colleen. 2006. "Venice's Vermin: 40,000 Pigeons." Associated Press, September 11. Retrieved May 2007 (http://www.msnbc.msn.com/id/14784143/).

Barwick, Sandra. 1996. "Fine for Thief Who Trapped 'Thousands' of Birds in Boxes." *Daily Telegraph*, July 2.

Bateson, Gregory. 1956. "The Message, 'This Is Play.'" In *Group Processes*, edited by B. Schaffner, 145–242. New York: Josiah Macy.

Bauman, Richard. 1972. "The La Have Island General Store: Sociability and Verbal Art in a Nova Scotia Community." *Journal of American Folklore* 85 (338): 330–43.

Beeley, Brian W. 1970. "The Turkish Village Coffeehouse as a Social Institution." *Geographical Review* 60:475–93.

Belguermi, Ahmed, Dalila Bovet, Anouck Pascal, Anne-Caroline Prévot-Julliard, Michel Saint Jalme, Lauriane Rat-Fischer, and Gérard Leboucher. 2011. "Pigeons Discriminate between Human Feeders." *Animal Cognition* 16 (6): 909–14.

Bell, Cyrus W. 1991. "Pigeons of Venice Rule the Roost." *Toronto Star*, January 5, F19.

Bell, Michael Mayerfeld. 1994. *Childerley: Nature and Morality in a Country Village*. Chicago: University of Chicago Press.

———. 1997. "The Ghosts of Place." *Theory and Society* 26:813–36.

Berger, John. 1980. "Why Look at Animals?" In *About Looking*, 3–30. New York: Vintage Books.

Birmingham Post. 1999. "Travel: Venice Pigeons Ruffle Feathers." February 6, 54.

Biro, Dora, David J. T. Sumpter, Jessica Meade, and Tim Guilford. 2006. "From Compromise to Leadership in Pigeon Homing." *Current Biology* 16:2123–28.

Blechman, Andrew D. 2006. *Pigeons*. New York: Grove Press.

Blumer, Herbert. 1969. *Symbolic Interactionism*. Berkeley: University of California Press.

Böcker, Anita. 2004. "The Impact of Host-Society Institutions on the Integration of Turkish Migrants in Germany and the Netherlands." Paper presented at the workshop on integration of migrants from Turkey in Austria, Germany, and the Netherlands, February 27–28, Bogaziçi University, Istanbul.

Bodio, Stephen. 1990. *Aloft: A Meditation on Pigeons and Pigeon-Flying*. New York: Lyons & Burford Publishers.

Bourdieu, Pierre. 1977. *Outline of a Theory of Practice*. Cambridge: Cambridge University Press.

Bourgois, Philippe. 1995. *In Search of Respect: Selling Crack in El Barrio*. Cambridge: Cambridge University Press.

Brown-Saracino, Japonica. 2011. "From the Lesbian Ghetto to the Ambient Community: The Perceived Costs and Benefits of Integration for Community." *Social Problems* 58 (3): 361–88.

Brubaker, Rogers. 1992. *Citizenship and Nationhood in France and Germany*. Cambridge, MA: Harvard University Press.

Brubaker, Rogers, Margit Feischmidt, Jon Fox, and Liana Grancea. 2006. *Nationalist Politics and Everyday Ethnicity in a Transylvanian Town*. Princeton, NJ: Princeton University Press.

Buber, Martin. 1970. *I and Thou*. Translated by W. Kaufman. New York: Touchstone.

Caldwell, Christopher. 2007. "Where Every Generation Is First Generation." *New York Times Magazine*, May 27, 44.

Čapek, Stella. 2010. "Foregrounding Nature: An Invitation to Think about Shifting Nature-City Boundaries." *City and Community* 9 (2): 208–24.

Catton, William R. Jr., and Riley E. Dunlap. 1980. "A New Ecological Paradigm for a Postexuberant Sociology." *American Behavioral Scientist* 24:15–47.

Colomy, Paul, and Robert Granfield. 2010. "Losing Samson: Nature, Crime, and Boundaries." *Sociological Quarterly* 51:355–83.

Cooley, Charles Horton. 1909. *Social Organization: A Study of the Larger Mind*. New York: Charles Scribner's Sons.

Cressey, Paul Goalby. (1932) 2008. *The Taxi Dance Hall: A Sociological Study in Commercialized Recreation and City Life*. Chicago: University of Chicago Press.

Cronon, William. 1995. *Uncommon Ground: Toward Reinventing Nature*. New York: Norton.

———. 1996. "The Trouble with Wilderness." *Environmental History* 1 (1): 7–28.

Csikszentmihalyi, Mihaly. 1975. *Beyond Boredom and Anxiety*. San Francisco: Jossey-Bass.

Daily Telegraph. 1965. "Lunchtime Crowd Has Festival Foretaste." September 16, 22.

Darwin, Charles. (1859) 1909. *The Origin of Species*. New York: P. F. Collier & Son.

Dave, Amish, Bistra Ivanova, and Ali Sutton. 2006. "The Challenge of Integration in the Berlin Educational System." New York: Humanity in Action. Retrieved September 2006 (http://www.migration-boell.de/web/integration/47_727.asp).

Davis, Robert C., and Garry R. Marvin. 2004. *Venice, the Tourist Maze: A Cultural Critique of the World's Most Touristed City*. Berkeley: University of California Press.

Desmond, Matthew. 2007. *On the Fireline: Living and Dying with Wildland Firefighters*. Chicago: University of Chicago Press.

Douglas, Mary. 1966. *Purity and Danger*. London: Penguin.

Duneier, Mitchell. 1994. *Slim's Table*. Chicago: University of Chicago Press.

———. 1999. *Sidewalk*. New York: Farrar, Straus, & Giroux.

Durkheim, Emile. (1912) 1995. *The Elementary Forms of Religious Life*. New York: Free Press.

Ehrkamp, Patricia. 2005. "Placing Identities: Transnational Practices and Local Attachments of Turkish Migrants in Germany." *Journal of Ethnic and Migration Studies* 31:345–64.

Evening Standard. 1997a. "MP Banks Fights 'Cruel' Plan to Starve Out Pigeons." January 17.

———. 1997b. "Plan to Starve Out the Pigeons Upsets Tourists." January 16.

Feuer, Alan. 2001. "They Don't Break for Statues: Racing Pigeons Hurry Home to Brooklyn, at 55 M.P.H." *New York Times*, October 16.

Fine, Gary Alan. 1997. "Naturework and the Taming of the Wild: The Problem of 'Overpick' in the Culture of Mushroomers." *Social Problems* 44 (1): 68–88.

———. 2003. *Morel Tales: The Culture of Mushrooming*. Chicago: University of Illinois Press.

Fine, Gary Alan, and Lazaros Christoforides. 1991. "Dirty Birds, Filthy Immigrants, and the English Sparrow War: Metaphorical Linkage in Constructing Social Problems." *Symbolic Interaction* 14:375–91.

Friedman, Thomas. 2005. *The World Is Flat: A Brief History of the Twenty-first Century*. New York: Farrar, Straus, & Giroux.

Fuchs, Stephan. 1989. "Second Thoughts on Emergent Interaction Orders." *Sociological Theory* 7 (1): 121–23.

Gans, Herbert. (1962) 1982. *The Urban Villagers: Group and Class in the Life of Italian-Americans*. New York: Free Press of Glencoe.

Garfinkel, Harold. 1967. *Studies in Ethnomethodology*. Englewood Cliffs, NJ: Prentice Hall.

Geertz, Clifford. 1963. *Old Societies and New States*. New York: Free Press.

———. 1973. *The Interpretation of Cultures: Selected Essays*. New York: Basic Books.

Gibson, James William. 2009. *A Reenchanted World: The Quest for a New Kinship with Nature*. New York: Holt.

Gieryn, Thomas F. 2000. "A Space for Place in Sociology." *Annual Review of Sociology* 26:463–96.

Gil-White, Francisco J. 1999. "How Thick Is Blood? The Plot Thickens . . . : If Ethnic Actors Are Primordialists, What Remains of the Circumstantialist/Primordialist Debate?" *Ethnic and Racial Studies* 22 (5): 789–820.

———. 2001. "Are Ethnic Groups Biological 'Species' to the Human Brain?: Essentialism in our Cognition of Some Social Categories." *Current Anthropology* 42:515–54.

Gmelch, George. 2001. *Inside Pitch: Life in Professional Baseball*. Washington, DC: Smithsonian Institution Press.

Goffman, Erving. 1959. *The Presentation of Self in Everyday Life*. New York: Anchor Books.

———. 1961. "Fun in Games." In *Encounters*, by E. Goffman, 15–81. Indianapolis, IN: Bobbs-Merrill.

———. 1963. *Behavior in Public Places*. New York: Free Press.

———. 1967. *Interaction Ritual: Essays on Face-to-Face Behavior*. Garden City, NY: Doubleday.

———. 1971. *Relations in Public*. New York: Harper & Row.

———. (1974) 1986. *Frame Analysis*. Boston: Northeastern University Press.

———. 1983. "The Interaction Order: American Sociological Association, 1982 Presidential Address." *American Sociological Review* 48 (1): 1–17.

Goode, David. 1994. *A World without Words*. Philadelphia: Temple University Press.

———. 2006. *Playing with My Dog, Katie: An Ethnomethodological Study of Canine-Human Interaction*. West Lafayette, IN: Purdue University Press.

Grasmuck, Sherri. 2005. *Protecting Home: Class, Race, and Masculinity in Boys' Baseball*. New Brunswick, NJ: Rutgers University Press.

Grazian, David. 2003. *Blue Chicago: The Search for Authenticity in Urban Blues Clubs*. Chicago: University of Chicago Press.

Greider, Thomas, and Lorraine Garkovich. 1994. "Landscapes: The Social Construction of Nature and the Environment." *Rural Sociology* 59 (1): 1–24.

Grosby, Steven. 1994. "The Verdict of History: The Inexpungable Tie of Primordiality—A Response to Eller and Coughlan." *Ethnic and Racial Studies* 17:164–71.

Gumbe, Andrew. 1995. "Hungry Pigeons Face Death in Venice." *Independent* (London), January 26, 9.

Haag-Wackernagel, Daniel, and H. Moch. 2004. "Health Hazards Posed by Feral Pigeons." *Journal of Infection* 48:307–13.

Hackney Gazette. 2006. "Pigeon Plop." October 9.

Harper, Douglas. 1987. *Working Knowledge: Skill and Craft in a Small Shop.* Chicago: University of Chicago Press.

Hendriks, Ruud. 1998. "Egg Timers, Human Values, and the Care of Autistic Youths." *Science, Technology, and Human Values* 23 (4): 399–424.

Herda-Rapp, Ann, and Theresa L. Goedeke, eds. 2005. *Mad about Wildlife.* Boston: Brill.

Hillman, Felicitas, and Hedwig Rudolph. 1997. "Redistributing the Cake? Ethnicisation Processes in the Berlin Food Sector." Working paper of the Wissenschaftszentrum Berlin für Sozialforschung.

Hönekopp, Elmar. 2003. "Non-Germans on the German Labour Market." *European Journal of Migration and Law* 5:69–97.

Hughes, Everett C. 1945. "Dilemmas and Contradictions of Status." *American Journal of Sociology* 50 (5): 353–59.

———. 1958. *Men and Their Work.* Glencoe, IL: Free Press.

Humphries, Courtney. 2008. *Superdove.* New York: Smithsonian Books.

Hunter, Albert. 1974. *Symbolic Communities: The Persistence and Change of Chicago's Local Communities.* Chicago: University of Chicago Press.

Irvine, Leslie. 2004. *If You Tame Me.* Philadelphia: Temple University Press.

Jacobs, Jane. 1961. *The Death and Life of Great American Cities.* New York: Modern Library.

Jenkins, Simon. 2001. No title. *Evening Standard,* January 25.

Jerolmack, Colin. 2007. "Animal Practices, Ethnicity and Community: The Turkish Pigeon Handlers of Berlin." *American Sociological Review* 72 (6): 874–94.

———. 2008. "How Pigeons Became Rats: The Cultural-Spatial Logic of Problem Animals." *Social Problems* 55 (1): 72–94.

———. 2009a. "Humans, Animals, and Play: Theorizing Interaction when Intersubjectivity Is Problematic." *Sociological Theory* 27 (4): 371–89.

———. 2009b. "Primary Groups and Cosmopolitan Ties: The Rooftop Pigeon Flyers of New York City." *Ethnography* 10 (4): 435–57.

Johnson, Angela. 1986. "The Branson Clean-up Brigade Go into Action." *Times,* August 22.

Johnston, Richard F., and Marián Janiga. 1995. *Feral Pigeons.* New York: Oxford University Press.

Katz, Jack. 1988. *The Seductions of Crime: Moral and Sensual Attractions in Doing Evil.* New York: Basic Books

———. 1997. "Ethnography's Warrants." *Sociological Methods and Research* 25 (4): 391–423.

———. 1999. *How Emotions Work*. Chicago: University of Chicago Press.

Kefalas, Maria. 2003. *Working-Class Heroes: Protecting Home, Community, and Nation in a Chicago Neighborhood*. Berkeley: University of California Press.

Kellert, Stephen R. 1993a. Introduction to *The Biophilia Hypothesis*, edited by S. R. Kellert and E. O. Wilson, 20–27. Washington, DC: Island Press.

———. 1993b. "The Biological Basis for Human Values of Nature." In *The Biophilia Hypothesis*, edited by S. R. Kellert and E. O. Wilson, 42–69. Washington, DC: Island Press.

Kellert, Stephen R., and Edward O. Wilson, eds. 1993. *The Biophilia Hypothesis*. Washington, DC: Island Press.

Kerr, J. 1985. "Pigeons Do the Dirty Work on Venice." *Telegraph*, May 28.

Kinney, Bob. 2005. "The Sun City Million #10." Million Dollar Pigeon Race. Retrieved in May 2006 (https://www.scmdpr.com/).

Kirshenblatt-Gimblett, Barbara. 1983. "The Future of Folklore Studies in America: The Urban Frontier." *Folklore Forum* 16 (2): 175–234.

Kligerman, Jack. 1978. *A Fancy for Pigeons*. New York: Hawthorn Books.

Kornblum, William. 1974. *Blue Collar Community*. Chicago: University of Chicago Press.

Kosut, Mary, and Lisa Jean Moore. Forthcoming. *Buzz: The Culture and Politics of Bees*. New York: NYU Press.

Lamont, Michèle. 2000. *The Dignity of Working Men: Morality and the Boundaries of Race, Class, and Immigration*. Cambridge, MA: Harvard University Press.

Lassiter, Unna, and Jennifer Wolch. 2005. "Changing Attitudes towards Animals among Chicanas and Latinas in Los Angeles." In *Land of Sunshine: The Environmental History of Metropolitan Los Angeles*, edited by G. Hise and W. Deverell, 267–87. Pittsburgh, PA: University of Pittsburgh Press.

Latour, Bruno. 1988. "Mixing Humans and Nonhumans Together: The Sociology of the Door Closer." *Social Problems* 35 (3): 298–310.

———. 1993. *We Have Never Been Modern*. Cambridge, MA: Harvard University Press.

Lee, Jooyoung. 2009. "Battlin' on the Corner: Techniques for Sustaining Play." *Social Problems* 56 (3): 578–98.

Levi, Wendell Mitchell. 1941. *The Pigeon*. Sumter, SC: Levi Publishing Co.

———. (1965) 1996. *Encyclopedia of Pigeon Breeds*. Sumter, SC: Levi Publishing Co.

Lévi-Strauss, Claude. 1963. *Totemism*. Boston: Beacon Press.

Liebow, Elliot. 1967. *Tally's Corner*. Boston: Little, Brown & Co.

Littlewood, Ian. 1991. *Venice: A Literary Companion*. London: John Murray.

Livingstone, Ken. 2001. "Why We Must Remove the Pigeons from Trafalgar Square." *Independent* (London), January 24.

Lloyd, Richard. 2005. *Neo-Bohemia: Art and Commerce in the Postindustrial City*. New York: Routledge.

Louv, Richard. 2005. *Last Child in the Woods*. Chapel Hill, NC: Algonquin Books.

Lydall, Ross. 2006a. "121 Dead in Trafalgar Sq; Hawks Are Killing Off Horrendous Numbers of Pigeons, Claim Campaigners, and They've Cost Us £215,000." *Evening Standard* (London), September 29, A22.

———. 2006b. "You're Only Meant to Scare the *?"?* Things!: Hawk Brought in by Ken to Frighten Off Pigeons Kills 14 Birds." *Evening Standard* (London), July 17, LSE15.

Mace, Rodney. 2005. *Trafalgar Square: Emblem of Empire*. London: Lawrence & Wishart.

Madden, Daniel M. 1964. "Venice Revisited." *New York Times*, May 17, XX39.

Malinowski, Bronislaw. 1948. *Magic, Science and Religion and Other Essays*. New York: Free Press.

Marcus, George E. 1995. "Ethnography in/of the World System: The Emergence of Multi-sited Ethnography." *Annual Review of Anthropology* 24:95–117.

Marzluff, John M., and Tony Angell. 2005. *In the Company of Crows and Ravens*. New Haven, CT: Yale University Press.

Mason, Sophie. 2000. "Mayor to Clean Up Trafalgar Square." *Building Design*, September 22, 4.

McCarthy, Michael. 2000. "Come and Have a Go if You Think You're Hard Enough: The Pigeons of London Send a Message to Ken Livingstone." *Independent* (London), October 14, 1–2.

McCullin, Don. 2005. *Trafalgar Square through the Camera*. London: National Portrait Gallery Publications.

Mead, George Herbert. (1934) 1967. *Mind, Self, and Society*. Chicago: University of Chicago Press.

Mechling, Jay. 1989. "'Banana Cannon' and Other Folk Traditions between Human and Nonhuman Animals." *Western Folklore* 48 (4): 312–23.

Miller, Stephen. 1973. "Ends, Means, and Gallumphing: Some Leitmotifs of Play." *American Anthropologist* 75 (1): 87–98.

Mitchell, Robert W., and Nicholas S. Thompson. 1986. "Deception in Play between Dogs and People." In *Deception*, edited by R. W. Mitchell and N. S. Thompson, 193–204. Albany: State University of New York.

———. 1990. "The Effects of Familiarity on Dog-Human Play." *Anthrozoös* 4 (1): 24–43.

Mooney, Jake. 2008. "Married, in a Way, to the Mob." *New York Times*, February 17, CY1.

Moore, Malcolm. 2007. "Venice to Halt Pigeon Feeding." *Daily Telegraph* (London), September 18, 18.

Morris, James. 1974. *The World of Venice*. New York: Harcourt Brace Jovanovich.

Morris, Jan. 1982. *A Venetian Bestiary*. New York: Thames & Hudson.

Mueller, Claus. 2006. "Integrating Turkish Communities: A German Dilemma." *Population Research and Policy Review* 25:419–41.

Muir, Hugh. 2006. "Hawks Do Their Worst but Cost of Pigeon War Is Problem for Mayor." *Guardian*, September 29.

Myerhoff, Barbara. 1980. *Number Our Days*. New York: Touchstone.

Myers, Gene. 1998. *Children and Animals*. Boulder, CO: Westview Press.

Nagy, Máté, Zsuzsa Ákos, Dora Biro, and Tamás Vicsek. 2010. "Hierarchical Group Dynamics in Pigeon Flocks." *Nature* 464:890–93.

Nash, Jeffrey E. 1989. "What's in a Face? The Social Character of the English Bulldog." *Qualitative Sociology* 12 (4): 357–70.

Nelson, Richard. 1993. "Searching for the Lost Arrow: Physical and Spiritual Ecology in the Hunter's World." In *The Biophilia Hypothesis*, edited by S. R. Kellert and E. O. Wilson, 201–28. Washington, DC: Island Press.

Nettleton, Philip. 2000. "Mayor Puts Square Pigeons to Flight." *Evening Standard* (London), October 2, 3.

New York Times. 1966. "Hoving Calls a Meeting to Plan for Restoration of Bryant Park." *New York Times*, June 22, 49.

Nicholls, Henry. 2009. "A Flight of Fancy." *Nature* 457 (12): 790–91.

O'Reilly, John. 1999. "That Sinking Feeling." *Observer*, July 18, 22.

Owen, Charles. 1989. "Battling for Trafalgar Square." *Country Life*, November 2, 164.

Owens, Erica. 2007. "Nonbiological Objects as Actors." *Symbolic Interaction* 30 (4): 567–84.

Park, Robert, and Ernest Burgess. 1925. *The City: Suggestions for Investigation of Human Behavior in the Urban Environment*. Chicago: University of Chicago Press.

Park, Robert Ezra. 1950. *Race and Culture*. Glencoe, IL: Free Press.

Pawson, Ernest. 1996. *Pigeon Racing: Science and Practice*. Pretoria, South Africa: J. L. van Schaik.

Peters, Wim. 2000. *Born to Win*. South Africa: Peters Publications.

———. 2005. "Challenge: Sun City Million Dollar Pigeon Race." *Racing Pigeon Digest*, April.

Philo, Chris. 1998. "Animals, Geography, and the City: Notes on Inclusions and Exclusions." In *Animal Geographies*, edited by J. Wolch and J. Emel, 51–71. New York: Routledge.

Philo, Chris, and Chris Wilbert, eds. 2000. *Animal Spaces, Beastly Places*. New York: Routledge.

Pinderhughes, Howard. 1997. *Race in the Hood: Conflict and Violence among Urban Youth*. Minneapolis: University of Minnesota Press.

Povoledo, Elisabetta. 2008. "Venice Bans Feeding in St. Mark's Square." *International Herald Tribune*, May 8.

Putnam, Robert. 2000. *Bowling Alone: The Collapse and Revival of American Community*. New York: Simon & Schuster.

Pyle, Robert Michael. 1993. *The Thunder Tree: Lessons from an Urban Wildland*. Boston: Houghton Mifflin.

Queens Tribune. 2006. "A Home for Diversity." Retrieved in January 2009 (http://www.queenstribune.com/guides/2005_PatchworkOfCultures/pages/Diversity.htm).

Raphael, Adam. 1967. "London's Problem Is No One's Pigeon." Retrieved from the Westminster Archives, June 14, 2006.

Rieder, Jonathan. 1985. *Canarsie: The Jews and Italians of Brooklyn against Liberalism*. Cambridge, MA: Harvard University Press.

Riesman, David, and Jeanne Watson. 1964. "The Sociability Project: A Chronicle of Frustration and Achievement." In *Sociologists at Work*, edited by P. E. Hammond, 235–321. New York: Basic Books.

Robins, Douglas M., Clinton R. Sanders, and Spencer E. Cahill. 1991. "Dogs and People: Pet-Facilitated Interaction in a Public Setting." *Journal of Contemporary Ethnography* 20 (1): 3–25.

Rozin, Paul, and Deborah Schiller. 1980. "The Nature and Acquisition of a Preference for Chili Pepper by Humans." *Motivation and Emotion* 4 (1): 77–101.

Sabloff, Annabelle. 2001. *Reordering the Natural World*. Toronto: University of Toronto Press.

Sallaz, Jeffrey J. 2009. *The Labor of Luck: Casino Capitalism in the United States and South Africa*. Berkeley: University of California Press.

Sanders, Clinton R. 1999. *Understanding Dogs: Living and Working with Canine Companions*. Philadelphia: Temple University Press.

———. 2003. "Actions Speak Louder than Words: Close Relationships between Humans and Nonhuman Animals." *Symbolic Interaction* 26 (3): 405–26.

Schmalenbach, Herman. (1927) 1977. *On Society and Experience*. Translated by G. Luschen and G. P. Stone. Chicago: University of Chicago Press.

Schwartz, Jane. 1986. *Caught*. New York: Ballantine Books.

Secord, James A. 1981. "Nature's Fancy: Charles Darwin and the Breeding of Pigeons." *Isis* 72:162–86.

Seligmann, Linda L. 2004. *Peruvian Street Lives: Culture, Power, and Economy among Market Women of Cuzco*. Chicago: University of Illinois Press.

Sennett, Richard. 2008. *The Craftsman*. New Haven, CT: Yale University Press.

Sewell, Brian. 2007. "Why I Shall Continue to Feed the Pigeons." *Evening Standard* (London), January 19, A13.

Sherman, Rachel. 2007. *Class Acts: Service and Inequality in Luxury Hotels*. Berkeley: University of California Press.

Shils, Edward. 1957. "Primordial, Personal, Sacred, and Civil Ties." *British Journal of Sociology* 8:130–45.

Simmel. Georg. 1949. "The Sociology of Sociability." *American Journal of Sociology* 55 (3): 254–61.

———. 1971. *On Individuality and Social Forms*. Edited by D. L. Levine. Chicago: University of Chicago Press.

Soeffner, Hans-Georg. 1997. *The Order of Rituals*. New Brunswick, NJ: Transaction Publishers.

Stewart, Phil. 2008. "Venice to Fine Tourists Who Feed Pigeons." Associated Press, April 30. Retrieved in May 2008 (http://news.yahoo.com/s/nm/20080430/lf_nm_life/venice_pigeons_dc).

Stringham, Sydney, Elisabeth Mulroy, Jinchuan Xing, David Record, Michael Guernsey, Jaclyn Aldenhoven, Edward Osborne, and Michael Shapiro. 2012. "Divergence, Convergence, and the Ancestry of Feral Populations in the Domestic Rock Pigeon." *Current Biology* 22:1–7.

Sullivan, Robert. 2010. "The Concrete Jungle." *New York Magazine*, September 12.

Suttles, Gerald D. 1968. *The Social Order of the Slum: Ethnicity and Territory in the Inner City.* Chicago: University of Chicago Press.

Taylor, Bill. 2007. "Picture and a Thousand Words: Consigned to Oblivion, London's Iconic Double-Decker Buses and the Conductors Who Manned Them Are Proving Surprisingly Difficult to Replace." *Toronto Star*, February 18, D02.

Tissot, Sylvie. 2011. "Of Dogs and Men: The Making of Spatial Boundaries in a Gentrifying Neighborhood." *City and Community* 10 (3): 265–84.

Tsing, Anna. 2005. *Friction: An Ethnography of Global Connection.* Princeton, NJ: Princeton University Press.

Tuan, Yi-Fu. 1984. *Dominance and Affection.* New Haven, CT: Yale University Press.

Turner, Mia. 2000. "China's Prized Pigeons Rule the Roost." *New York Times*, January 12.

Turner, Victor. 1974. *Dramas, Fields, and Metaphors: Symbolic Action in Human Society.* Ithaca, NY: Cornell University Press.

Urry, John. 1990. *The Tourist Gaze: Leisure and Travel in Contemporary Societies.* New York: Sage Publications.

Veniceword.com. 2007. "Pigeons: A Problem in Venice." Retrieved May 2007 (http://www.veniceword.com/news/58/pigeons.html).

Viddich, Arthur J., and Joseph Bensman. 1958. *Small Town in Mass Society: Class, Power, and Religion in a Rural Community.* Princeton, NJ: Princeton University Press.

Vitale, Alex. 2009. *City of Disorder: How the Quality of Life Campaign Transformed New York Politics.* New York: NYU Press.

Wacquant, Loïc. 2004. *Body and Soul: Notebooks of an Apprentice Boxer.* Oxford: Oxford University Press.

Waters, Mary C. 1990. *Ethnic Options: Choosing Identities in America.* Berkeley: University of California Press.

Watson, Jeanne, and Robert J. Potter. 1962. "An Analytic Unit for the Study of Interaction." *Human Relations* 15 (3): 245–63.

Weber, Jacqueline, Daniel Haag, and Heinz Durrer. 1994. "Interaction between Humans and Pigeons." *Anthrozoös* 7 (1): 55–59.

Weber, Max. (1918) 1946. *From Max Weber: Essays in Sociology.* Translated by H. H. Gerth and C. Wright Mills. Oxford: Oxford University Press.

———. (1922) 1978. *Economy and Society.* Berkeley: University of California Press.

Wellman, Barry. 1979. "The Community Question: The Intimate Networks of East Yorkers." *American Journal of Sociology* 84 (5): 1201–31.

West End Extra. 2000. "Pigeons Get Reprieve." December 15.

Whyte, William Foote. (1943) 1993. *Street Corner Society: The Social Structure of an Italian Slum.* Chicago: University of Chicago Press.

Whyte, William H. 1988. *City: Rediscovering the Center.* New York: Anchor Books.

Willey, David. 1995. "It's Sink or Swim for Venetian Pigeon Population." *Gazette* (Montreal), December 23, G7.

Williams, Raymond. 1973. *The Country and the City.* Oxford: Oxford University Press.

Willis, Paul E. 1981. *Learning to Labor: How Working Class Kids Get Working Class Jobs*. New York: Columbia University Press.
Wilson, Edward O. 1993. "Biophilia and the Conservation Ethic." In *The Biophilia Hypothesis*, edited by S. R. Kellert and E. O. Wilson, 31–41. Washington, DC: Island Press.
Wilson, William J., and Richard P. Taub. 2007. *There Goes the Neighborhood*. New York: Vintage Books.
Wirth, Louis. 1938. "Urbanism as a Way of Life." *American Journal of Sociology* 44 (1): 1–24.
Wolch, Jennifer. 1998. "Zoöpolis." In *Animal Geographies*, edited by J. Wolch and J. Emel, 119–38. New York: Verso.
Wolch, Jennifer, and Jody Emel, eds. 1998. *Animal Geographies*. New York: Verso.
Wynn, Jonathan R. 2011. *The Tour Guide: Walking and Talking New York*. Chicago: University of Chicago Press.
Yakin, Heather. 2008. "Pigeon Racing a Growing Sport." *Times Herald–Record*, April 10.
Young, Michael, and Peter Willmott. (1957) 1992. *Family and Kinship in East London*. Berkeley: University of California Press.
Zerubavel, Eviatar. 1997. *Social Mindscapes: An Invitation to Cognitive Sociology*. Cambridge, MA: Harvard University Press.
Zorbaugh, Harvey. 1929. *The Gold Coast and the Slum: A Sociological Study of Chicago's Near North Side*. Chicago: University of Chicago Press.
Zukin, Sharon. 1989. *Loft Living: Culture and Capital in Urban Change*. New Brunswick, NJ: Rutgers University Press.

INDEX

Numbers in italics indicate photographs.

Above Brooklyn (film), 221–22
abusive joviality, 139, 171, 186
African American pigeon fanciers, 91–96
alienation 126, 219
Allay, Fabian, 206
ambivalence, 15
Anderson, Elijah, 136, 148
Angelo, Hilary, 231
animal rights, 65
animals, social experience of, 14, 72, 104–5, 225–26. *See also* dogs; ducks; geese; hawks; pigeons
Anna ("the pigeon lady"), 26, 28–30, 42
antiurbanism, 125
apartheid, 195–96, 203, 205, 215
Appadurai, Arjun, 19
asocial nature, 13, 14, 18, 230, 232, 235–37
asymmetrical interaction, 37–40
attachment, unseemly, 140–43
auctioning pigeons, 195, 199, 210–11
authenticity, 14, 53, 125, 221, 230, 233

ballbusting. *See* abusive joviality
Banks, Tony, 59, 60
basketing, 169–74, 182, 190, 199–201
beauty, 145

beekeeping, 15, 228, 230, 231, 233
Bell, Michael, 45, 54, 125, 232
Belli, Ali, 118
Berger, John, 2, 14
Berlin, Turkish immigrants in, 6, 109–32
betting, 171, 174, 176, 192
Bianchi, Franco, 159–60, 171–72, 174, 182–86, 188
biodiversity, 236
biologists, 67
biophilia, 12, 41, 103–4; as distinctively Turkish, 121, 127
Birds, The (film), 48
Bishop, Dave, 62–64, 68, *69*
block associations, 24
Bond, James, 50
boundary work, 126
Broadway Pigeons and Pet Supplies, 79, 133–65
Bronx Homing Pigeon Club, 6, 17, 159–91, *163*, 219–20. *See also* pigeon racing
Brooklyn, NY, racial and ethnic composition, 79, 133, 150–52, 162
Browning, Elizabeth Barrett, 47
Buber, Martin, 39

Çabakçor, Ahmet, 109, *111*, 115, 120
Cacciari, Massimo, 55, 71, 73
Cesur, Timur, 117–18, 119–20
Chicago, 225, 228–29, 231
Chicago school of sociology, 129
chicken coops, 15, 228, 231, 233
childhood experience with pigeons, 114, 115, 131–32
children, 2, 14, 41, 69, 126
China, 210
Çim, Asim, 115–16
citizenship, 250n4
class, social and economic, 80, 99–103, 133, 228–31, 252n6. *See also* gentrification
clocking, 177–88, 192–93, 201–10, 220
clubs, 93, 196, 205. *See also* Bronx Homing Pigeon Club
cockfighting, 152, 161, 174, 190, 215
commercialization, 216
"Community Lost" hypothesis, 12, 229–30. *See also* "Nature Lost" hypothesis
competitive sociability, 136–39, 148, 171
Conradie, Rob, 199, 204, 212
Cooley, Charles H., 134
cosmopolitan ties, 134, 152–54
Coston, Cecil, 188–89
craftwork, 216, 218–89
Cronon, William, 13, 72, 232, 235–36
cruelty to animals, 60, 70, 152

Darwin, Charles, vi, 10, 113, 145
Dede, Ahmet, 115, 126
deep play, 161
deer, 15
Dallacco, Henry, 180–81
Demir, Turan, 121
discrimination, 129. *See also* apartheid
disease, 16, 103; incorrectly associated with pigeons, 8–9, 47, 57, 72, 227
disenchantment, 18, 189–90, 228
disgust, 52–53
dogfighting, 152

dog racing, 180
dogs, 15, 31, 37, 39, 72, 105, 121, 231
doping, 182, 199, 220
Douglas, Mary, 3, 72, 73
ducks, 64
Durkheim, Emile, 14, 127, 232
Durkut, Israfil, 114

elderly people, 41, 66. *See also* marginalization
enchantment, 12–13, 53, 125, 162, 195, 210, 217–20, 228. *See also* magic
Engelbrecht, Tersia, 201–3, *203*
Engelbrecht, William "Butch," 204
environmentalism, 127, 235, 237
equality of opportunity, 18, 212, 214
ethnicity, 150–53, 225; pigeon keeping as reinforcer of, 6, 111, 125–26, 233
eviction, 224–25
expertise, 101–3

falconry, 44, 60, 62–64, 67
family, 124, 126, 135, 154, 155, 181, 182, 183, 189
Father Demo Square, 1–3, *2*, 5, 17, 23, *24*, 233
feces, pigeon, 1, 2, 52, 59; cleanup, 80, 87, 93, 99, 183, 106; damage caused by, 8, 47, 57, 71, 227
fertility, pigeons as symbol of, 9
Flight, American Domestic (breed of pigeon), 83–85, 146–47
flock interaction, 97
flow, 38
foreignness, 130
Friedman, Thomas, 215

gambling, casino, 196
Gangone, Carmine, 3–5, *4*, 79–89, *86*, *88*, 108, 133, 136–37, 141, 152, 153, 221
Garcia, Richie, 96–98, *98*, 150–51

Geertz, Clifford, 15, 127, 161, 174, 190, 194, 215, 234
geese, 64, 230
gender, 142, 155–56, 201–4, 215
gentrification, 135, 221–22, 224, 251n1
ghosts, 45, 74
Gibson, James, 12, 14, 104, 125, 127, 229–30, 236
globalization, 18, 19, 204–5, 212, 215, 235
Goffman, Erving, 17, 31, 40, 137, 234
government: New York City, 3, 6, 8, 24, 224; London, 9, 45–46, 58–60, 68–73; Venice, 46, 55–56, 71, 73
Grasmuck, Sherry, 154
green space, 15

Harper, Douglas, 101, 139
hawks, 13, 93, 103, 119, 141, 144, 167; celebrity, 15, 104, 229; trained, 44, 60, 62–64, 67
heritage wardens, 68, 74
Hilton, Paris, 60
homeless people, 3, 24, 25, 27–28, 35, 40, 41, 72–73, 227
homesickness of migrants, 110
homosexuality, 72
horses, 121
Hoving, Thomas P., 72

immigrants: in New York, 4, 79, 88; in Venice, 50; Turkish in Berlin, 6, 109–32, *124*; possibility of return, 126; recreational options, 122; unemployment amongst, 112, 117, 131
invasive species, 228
Irvine, Leslie, 37, 104, 106
Italian Americans, 4; neighborhood identity, 5, 79

Jacobs, Jane, 1, 23, 25, 40, 42, 237
Joey's. *See* Broadway Pigeons and Pet Supplies

Katz, Jack, 137
Kefalas, Maria, 164
Kellert, Stephen, 104
Kirshenblatt-Gimblett, Barbara, 222–23
Kligerman, Jack, 151
Kornblum, William, 152
Krog, Vyver, 208, *211*

labor, division of, 194, 216, 235
Latour, Bruno, 14
Levi, Wendell, 84
Lévi-Strauss, Claude, 14, 127, 232
Livingstone, Ken, 9, 45–46, 59, 67, 70–71, 73
London. *See* Livingstone, Ken; Trafalgar Square
Louv, Richard. *See* nature deficit disorder
lying, 178–79

magic, 53, 177. *See also* enchantment
Malinowski, Bronislaw, 190
manhood, 118, 126, 133, 138, 142, 150, 155–56, 223; and expertise, 139–43
Marcus, George, 19
marginalization, 41, 126, 129–30, 250n4
Mary Poppins, 59, 66
McGuinness, Marty, 165–69, *168*, 173, 175, 187–88, 220, 222–24
MDPR. *See* Million Dollar Pigeon Race
Mead, George Herbert, 5, 106, 234
meritocracy, 194, 212, 215
messenger pigeons, 162
Meyer, Zandy, 196–97, 204, 213
Million Dollar Pigeon Race, 18, 192–220, 235; compared to New York racing, 209–10; economics of, 197–98, 211–12, 214
Mitchell, Robert, 37
mobility, social, 126
Monaco, Sal, 101–12, 106, 154
money, 144, 193–94, 216. *See also* pigeon keeping, costs of; pigeon racing, costs of

Morris, Jan, 47
Muir, John, 13
Musto, Joe, 169, 171–72, 189
Musto, Vinny, 169, *170*, 172–76, 178–81

Nash, Jeffrey, 106
nature, free of human interference, 13, 14, 18, 72, 230, 232–33, 235–37
nature, illusion of control over, 176
nature as antagonist, 161, 172
nature deficit disorder, 12
"Nature Lost" hypothesis, 12, 225–29, 232–33
nature worship, 127
neighbors, 148, 224
neoliberalism, 18, 194, 215
nonchalance, as marker of masculinity, 134, 140, 142–43, 146
nostalgia, 13, 110, 127, 148, 223

"Panama" (Delroy Sampson), 91–96, *94*
Park, Robert, 149
parks, public, 15; New York, 5
partners in pigeon keeping, 81, 151
past, connection with, 5, 10, 80, 89, 121, 125–26; disjunction from, 71
Penkin, Saville, 212
perceptual socialization, 145
Peters, Theo, 195
Peters, Wim, 193–94, 195, *203*, 204, 207, 217, 254n26
pets, 106–7, 175
pet shops, 3, 79, 89, 195, 222. *See also* Broadway Pigeons and Pet Supplies
Philo, Chris, 15, 72, 226
photography, normalizing behavior, 47, 48–49, 51, 57, 64
Piazza San Marco, 44–56, *49*, *51*, *54*, 74, 233; compared to Father Demo Square, 3, 17
pigeons: breeding, 105, 149, 165, 167; breeds, significance of, 83–85; as connection to nature, 110, 119; dependence on people, 75; feral ("street"), 16–17, 145, 159; as homeless species, 73, 22; homing ability of, 164–65; ideal, 145, 175; manipulating humans, 25, 29–30, 38, 66; as marker of disorder, 2–3, 67, 70–72; naming, 182–83, 191; natural history of, 7–8, 9, 11, 12, 226; as pests, 8, 41; stray, 87, 97–98; value of, 144, 175; Venetian, legendary history of, 46–47
pigeon feeding, 1–2, 6, 17, 24–27, 40–41; banned, 7, 23, 27, 44–45, 55–56, 58–60, 68–71; as cultural attraction, 45, 47–48, 57, 69–70, 236; as deviance, 45, 52; encouraged, 11, 48–50; focused, 31–37; and interpersonal interaction, 27, 42; normalized, 52; as ritual, 7, 17, 44, 49, 51, 66, 75; rogue, 65, 68, 70, 72; sanctioned, 45, 62; unfocused, 27–31. *See also* Piazza San Marco; Trafalgar Square
pigeon keeping: banned, 225; in Berlin, 6, 17, 109–32; costs of, 131, 135, 144, 147, 164, 182, 192; in New York, 3–5, 17, 79–108, 219
pigeon racing, 7, 17–18, 235; costs of, 182, 192, 197, 205, 212, 217; hazards of, 159, 172, 176, 181 (*see also* hawks) ; history of, 10, 162; preparation for, 117, 143, 161, 165–66, 173, 192, 198–99, 219; rewards of, 159, 174, 180, 188, 197, 208, 217; scoring, 164, 179; seasonality of, 160–61, 209; seduction of, 188–91. *See also* basketing; Bronx Homing Pigeon Club; clocking; Million Dollar Pigeon Race
pigeon wars, 99
prediction, 175
primordial ties, 128–29, 133, 152–53, 234; cross-species, 132
privacy, 40

professionalization, 194, 212. *See also* rationalization
progress, social, 206
property damage by pigeons, 8–9
public health, 8–9
public space, 23, 27, 44; and social vulnerability, 40. *See also* Father Demo Square; parks, public; Piazza San Marco; Trafalgar Square
Pyle, Robert Michael, 12

quality of life, tied to pigeons, 3, 58–59, 224, 227
Queen Elizabeth II, 213
quotation scheme, 20

raccoons, 39
race, 146, 150–53, 164, 203–4, 216, 251n1; transcended for primary group attachments, 6, 152–54. *See also* ethnicity
racial mixing, 133–34
racing. *See* pigeon racing
rationalization: of pigeon racing, 195, 201, 216–17, 219, 234; of society, 13
"rats" (fanciers' name for street pigeons), 87, 145, 146
"rats with wings," 7, 8, 45, 71, 72, 227–28; invention of usage, 72
Rayner, Bernie, 60, 65–66, 70
reputation, 149, 167. *See also* status
risk, 144, 175; constrained, 190
ritual, 126, 161, 177, 190, 201. *See also* pigeon feeding, as ritual

Sallaz, Jeffrey, 196
Sampson, Delroy "Panama," 91–96, *94*
San, Sinan Samil, 120
Sanders, Clinton, 37, 106
Scott, Joey, 89–91. *See also* Broadway Pigeons and Pet Supplies
Scott, Michael, 139–40
scripting, social 50

Secord, James, 10
seed (pigeon food) vendors, in London, 56, 60, 65, 66, 70, 71, 247n26; in Venice, 44, 46, 48–51, *51*
self, social, 106, 136–37, 140, 149
Seligmann, Linda, 20
Shils, Edward, 129
sidewalk life, 23, 25, 40, 236
Simmel, Georg, 25, 39
Smith, Paul, 199–200, 213
sociability, 25. *See also* competitive sociability
Soeffner, Hans-Georg, 102
solidarity, 128, 134, 149, 153, 161, 169, 172, 191, 232
solitude, 41
space, 45. *See also* public space
status, 5, 139, 190, 194, 214, 215, 233, 236
Steenkamp, Willie, 198–99
Stiglingh, Henry, *203*
Stiglingh, Petra, 201–2, *203*
Sullivan, Robert, 236
Sun City. *See* Million Dollar Pigeon Race
surveillance, 74
symbolic interaction, 37, 234

taboo, 45, 52, 71, 73, 226, 228, 234
teahouses, Turkish in Berlin, 122–23
Thompson, Nicholas, 37
Thoreau, Henry David, 4, 13
tiplet (breed of pigeon), 84–85, 97, 147–48
Tissot, Sylvie, 231
totemism, 13–14, 104, 127–28, 232
tough masculinity, 154
tourism, 71, 213. *See also* Piazza San Marco; Trafalgar Square
tourist gaze, 52
tourists, 23, 25, 44, 47–51, 55, 66–67
Trafalgar Square, 17, 44, 45, 56–70, *58*, 74; policing, 68, 74; site of cultural events, 61–62, 64, 67
trust, 34, 48, 203

Tuan, Yi-Fu, 15, 105–6
tumblers, Turkish (breed of pigeon), 109–10, 113–14, 120; as symbol of Turkish identity, 128
Turkey, culture of, 120; pigeon keeping in, 121. *See also* immigrants, Turkish in Berlin
Tyson, Mike, 80, 164

Uluagaç, Hasan, 121, 123–24
Ünver, Mehmet, 115, 123, 126, *131*
Up on the Roof (film) 255n7
urban homesteading, 228–29
urbanism, 5, 16, 237
urbanization, 12, 14–15, 226
Urry, John, 52

Venice, 46. *See also* government, Venice; Piazza San Marco
Viola, Frank, 160
violence, 141–43
vulnerability, in public spaces, 40

Waters, Mary, 132
Weber, Max, 13, 17, 189–90
Wilbert, Chris
Wilson, Edward O. *See* biophilia
women, 201–4, 216
work, 100–103, 166–67, 172
Wynn, Jonathan, 53

Yasaroglu, Abit and Abdi, 116–18, *117*
Yildirim, Beyhan, 109, *111*